NATURAL DISASTER RESEARCH, PREDICTION AND MITIGATION

HURRICANE DAMAGE: IMPLICATIONS OF INCREASED RISK ON THE FEDERAL BUDGET

NATURAL DISASTER RESEARCH, PREDICTION AND MITIGATION

Additional books in this series can be found on Nova's website under the Series tab.

Additional e-books in this series can be found on Nova's website under the e-book tab.

NATURAL DISASTER RESEARCH, PREDICTION AND MITIGATION

HURRICANE DAMAGE: IMPLICATIONS OF INCREASED RISK ON THE FEDERAL BUDGET

MARTHA ATKINSON
EDITOR

nova publishers
New York

Library of Congress Cataloging-in-Publication Data

ISBN: 978-1-53610-035-8

Published by Nova Science Publishers, Inc. † New York

Contents

PREFACE

Damage from hurricanes is expected to increase significantly in the coming decades because of the effects of climate change and coastal development. In turn, potential requests for federal relief and recovery efforts will increase as well. The Congressional Budget Office has estimated the magnitude of the increases in hurricane damage and the associated amounts of federal aid if historical patterns hold. This book provides a detailed discussion on the implications of increased risk on the federal budget of hurricane damage in the upcoming years.

In: Hurricane Damage
Editor: Martha Atkinson
ISBN: 978-1-53610-035-8

Chapter 1

POTENTIAL INCREASES IN HURRICANE DAMAGE IN THE UNITED STATES: IMPLICATIONS FOR THE FEDERAL BUDGET*

Terry Dinan, with contributions from Tristan Hanon and Jon Sperl

NOTES

Unless otherwise indicated, all monetary values are expressed in 2015 dollars and all years are calendar years.

Percentages in tables and figures may not add to 100 because of rounding.

SUMMARY

Damage from hurricanes is expected to increase significantly in the coming decades because of the effects of climate change and coastal development. In turn, potential requests for federal relief and recovery efforts will increase as well. The Congressional Budget Office has estimated the

* This is an edited, reformatted and augmented version of a report issued by the Congressional Budget Office, June 2016.

magnitude of the increases in hurricane damage and the associated amounts of federal aid if historical patterns hold. In addition, CBO examined three approaches to reducing the amount of such federal assistance: limiting greenhouse gas emissions; shifting more costs to state and local governments and private entities, thereby reducing coastal development; and investing in structural changes to reduce vulnerability to hurricanes. The accompanying working paper provides a detailed discussion of the data and methodology CBO used to estimate hurricane damage.[1]

What Are CBO's Estimates of Hurricane Damage and of Related Federal Spending?

CBO concludes that, over time, the costs associated with hurricane damage will increase more rapidly than the economy will grow. Consequently, hurricane damage will rise as a share of gross domestic product (GDP), which provides a measure of the nation's ability to pay for that damage. According to the agency's estimates, expected annual damage currently amounts to 0.16 percent of GDP (or about $28 billion); by 2075, however, that figure reaches 0.22 percent (equivalent to about $39 billion in today's economy; see Summary Figure 1). Roughly 45 percent of that increase is attributable to climate change and 55 percent to coastal development.

The percentage of the population exposed to substantial damage is likely to grow as well. In CBO's estimation, less than 0.4 percent of the U.S. population, or about 1.2 million people, currently lives in counties where expected hurricane damage per capita is greater than 5 percent of the county's average per capita income. By 2075, that share will rise to 2.1 percent of the population, or about 10 million people, CBO estimates.

In its analysis, CBO estimated annual federal spending for relief and recovery as a percentage of expected hurricane damage. If that percentage stays roughly the same as it has been over the past decade—a prospect referred to in this report as a historical cost scenario—it will rise from 0.10 percent of GDP under current conditions (equal to $18 billion) to 0.13 percent of GDP in 2075 (about $24 billion in today's economy).[2] If federal spending as a percentage of hurricane damage changed, those amounts could be larger or smaller.

How Did CBO Estimate Hurricane Damage?

CBO estimated the change in damage from hurricanes by comparing expected damage under current conditions with expected damage in selected future years—2025, 2050, and 2075—under the conditions that are expected to prevail at the time. Expected hurricane damage in any given year will depend on four conditions:

- Sea levels in different states,
- The frequency of hurricanes of various intensities,
- The population in coastal areas, and
- Per capita income in coastal areas.

For each set of conditions, CBO estimated expected damage using commercially developed, state-of-the-art "damage functions" (which translate hurricane occurrences, state-specific sea levels, and current property exposure into state-specific expected damage) and the agency's own assessment of the relationship between changes in population and per capita income and changes in hurricane damage.

Two of the four conditions—sea levels and the frequency of hurricanes—are affected by climate change. Strong consensus exists within the scientific community that climate change, the result of growing emissions of greenhouse gases worldwide, will cause sea levels to rise, leading to more-destructive storm surges. The effect of climate change on hurricanes is less certain, but scientists find that it could increase the frequency of hurricanes in the North Atlantic, particularly the most intense categories of hurricanes.

The other two conditions—population and per capita income in counties that are vulnerable to damage from hurricanes—are affected by coastal development. Ongoing trends in coastal development will similarly exacerbate hurricane damage, even in the absence of any increase in sea levels or in the frequency of hurricanes.

In its analysis, CBO used projections made by leading researchers to construct distributions (indicating the range and probability of alternative outcomes) of future sea levels and the frequency of hurricanes. The agency used its own projections to construct distributions of population and per capita income. On the basis of those projections, CBO constructed a distribution of expected damage for each future year considered in this report by conducting thousands of simulations. Each simulation included a unique set of draws (random selections) from the distributions of the four underlying conditions and yielded an estimate of expected damage based on those draws.

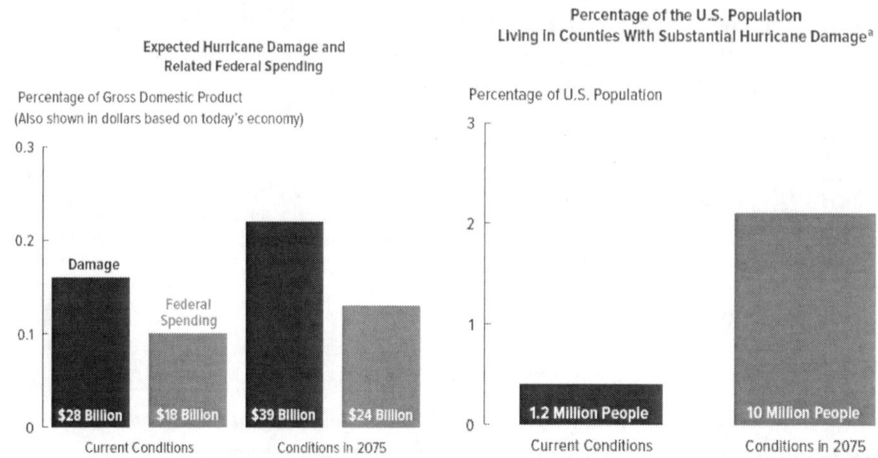

Source: Congressional Budget Office.

CBO's estimate of expected annual hurricane damage at present—the "reference case"—is based on estimates of the current frequency of hurricanes, current state-specific sea levels, and the current valuation of property exposed to hurricanes.

CBO's estimates of federal spending are based on a scenario in which federal aid for relief and recovery—measured as a percentage of the damage resulting from hurricanes—stays roughly the same as it has been over the past decade.

CBO's estimates of expected hurricane damage in 2075 are based on the average results of 5,000 simulations, with each simulation using a unique set of draws (random selections) for the underlying conditions that determine expected damage.

[a] People exposed to substantial hurricane damage are defined as those living in counties in which per capita expected damage is greater than 5 percent of the county's per capita income.

Summary Figure 1. Estimates of Hurricane Damage, Federal Spending, and the Substantially Affected Population.

How Did CBO Estimate Future Federal Spending Related to Hurricane Damage?

Federal aid that is provided following hurricanes supports emergency relief operations, long-term recovery activities, and a variety of programs that are designed to improve the resiliency of infrastructure and to prepare communities for future disasters. Most such spending is not mandated by law; rather, it is the outcome of decisions made by policymakers in the aftermath of disasters and is funded primarily through supplemental appropriations.

Federal spending in response to hurricanes varies from storm to storm. However, measured as a percentage of total damage—estimates of which are produced by the National Oceanic and Atmospheric Administration (NOAA)—such spending has averaged about 60 percent for the nine hurricanes that made landfall since August 2005, when Hurricane Katrina struck the Gulf Coast. For CBO's historical cost scenario, the agency estimates that federal expenditures would continue to average 60 percent of total damage from hurricanes.

What Policies Might Decrease the Pressure for Federal Spending in the Aftermath of Hurricanes?

In considering how to ease the pressure to spend federal dollars on relief and recovery from hurricane damage, CBO examined three diverse approaches.

Limit Greenhouse Gas Emissions

A coordinated global effort to significantly reduce emissions could lessen hurricane damage in the United States between now and 2075, but the extent of the reduction would be uncertain and it would probably occur in the latter half of this century because the rise in sea levels has already been set in motion and would be hard to slow down. However, a significant reduction in U.S. greenhouse gas emissions, without corresponding decreases in the emissions of other large economies, would probably not reduce hurricane damage appreciably between now and 2075, in part because U.S. emissions constitute a shrinking share of global emissions.

Shift More Costs to State and Local Governments and Private Entities, Thereby Reducing Coastal Development

CBO projects that, continuing historical trends, the population in coastal areas will grow more rapidly than in the United States as a whole. To the extent that households, businesses, and state and local governments in coastal areas do not bear the full cost of hurricane damage, such growth is subsidized by U.S. taxpayers in general. Boosting the share borne by private and public entities at the state and local levels would give people an incentive to more fully account for expected hurricane damage when choosing where to live and locate businesses, thereby reducing development in vulnerable areas. Policies that would accomplish those goals include the following:

- Expanding the use of flood insurance and raising premiums to more accurately reflect the costs of expected damage;
- Increasing the minimum amount of statewide per capita damage that is the primary consideration for providing federal assistance; and
- Reducing the share of costs borne by the federal government (as opposed to state and local governments) for assistance provided from the Federal Emergency Management Agency's (FEMA's) Disaster Relief Fund (DRF), which accounts for roughly one-half of federal spending on relief and recovery from hurricanes.

Invest in Structural Changes to Reduce Vulnerability to Hurricanes

In recent years, federal agencies have placed a greater emphasis on measures designed to reduce vulnerability to future hurricane damage, such as elevating roads and using flood-resistant building materials. Such hazard-mitigation measures typically increase the up-front costs of construction or restoration but reduce costs associated with future damage. To the extent that up-front investment pays off, the federal government could reduce its hurricane-related spending by undertaking more hazard mitigation or by providing incentives for individuals, businesses, and state and local governments to do so.

Overview of Climate Change and Coastal Development and Their Effects on Hurricane Damage

The magnitude of hurricane damage in future years will be affected both by climate change and by coastal development. Each individual factor is likely to increase the amount of damage; moreover, each factor will amplify additional damage caused by the other. For example, rising sea levels will lead to stronger storm surges, which will compound the additional damage that will occur if more homes and businesses are built in vulnerable coastal areas.

Climate Change

Human activities around the world—primarily the burning of fossil fuels and widespread changes in land use—are producing growing emissions of greenhouse gases. Experts in the scientific community have concluded that a portion of those emissions are absorbed by the oceans, but a substantial fraction persists in the atmosphere for centuries, trapping heat and warming the Earth's atmosphere.

Accumulating heat in the atmosphere has disrupted the natural balance that is normally maintained between heat in the atmosphere and heat in the oceans. The oceans respond to that imbalance by absorbing heat from the atmosphere; however, that absorption occurs very slowly. As a result, even if greenhouse gas emissions were to cease, the oceans would continue to become warmer for centuries until they eventually came into balance with the atmosphere. Continued warming of the oceans will cause sea levels to rise through two processes: First, it will cause oceans to expand because the volume of water increases as it warms; second, in concert with the warming of the atmosphere, it will trigger melting of sections of the ice sheets in Greenland and Antarctica.

CBO's estimates of hurricane damage in selected future years—2025, 2050, and 2075—are based on experts' assessments of the degree to which climate change will alter sea levels along the U.S. coastline and affect the frequency of hurricanes (the average number of occurrences in a given year). Although scientists generally agree that climate change will cause sea levels to rise, the pace at which that will occur is uncertain. Considerably less agreement exists about the effect of climate change on the frequency of hurricanes of various strengths. (There are five categories of hurricanes, which are classified on the basis of their wind speed. Category 5 storms are the most intense.) Effects on hurricane frequency are complicated by the fact that scientists expect climate change to increase some factors, such as sea surface temperatures, that increase the likelihood of hurricanes forming, while also increasing other factors, such as wind shear, that decrease that likelihood. Models produce a wide range of potential outcomes, but many predict an increase in the probability of major hurricanes (Category 3 and higher) in the North Atlantic. (That possibility is discussed in greater detail below.)

Coastal Development

Between 2000 and 2010, the most recent decade for which census data are available, the population of counties that are susceptible to hurricane damage grew 22 percent faster than the overall U.S. population. (Even more rapid growth occurred between 1950 and 2000, when the population of coastal counties grew over three times faster than that of the United States. That growth was fueled, in part, by the increased availability of air-conditioning and is unlikely to be repeated.)

Continued development along the East and Gulf coasts of the United States is likely to amplify hurricane damage—even in the absence of climate change—simply by putting more people and property in harm's way. Moreover, rising sea levels and changes in the frequency of hurricanes will compound the growth in damage caused by coastal development. CBO approximated increases in property exposure by estimating increases in population and per capita income in areas that are vulnerable to hurricane damage. The agency's analysis also accounted for the possibility that substantial increases in hurricane damage could slow the rate of coastal development. Finally, CBO considered alternative estimates of the relationship between increases in exposed property and increases in hurricane damage, accounting, for example, for the possibility that taller buildings could lessen per capita damage from storm surges.

PROJECTING HURRICANE DAMAGE: CBO'S METHODOLOGY

Actual hurricane damage in any given future year could vary enormously for several reasons. First, hurricanes— particularly the most intense hurricanes—occur infrequently and irregularly. Second, the magnitude of damage caused by a hurricane of a particular strength, such as a Category 3 storm, could vary greatly depending on where it makes landfall. The same category of hurricane could result in minor damage if it struck an unpopulated section of the coast or major damage if it struck a large city. Finally, the timing of impact can be important. A hurricane that makes landfall at high tide, as Hurricane Sandy did in 2012, will result in far more damage than if the same hurricane hit at low tide. In combination, those factors make it difficult to discern changes in long-term trends in hurricane damage from changes in actual damage over time.

Box 1. Expected Versus Actual Hurricane Damage in a Given Year

Expected damage reflects the average annual frequency of hurricanes and the average annual damage that a particular category of hurricane would impose for a given sea level and amount of coastal development. That average annual damage accounts for all of the paths that a hurricane of a given category could follow—for instance, whether it heads harmlessly out to sea or makes landfall at different points along the coast of the United States.

Expected damage includes the potential for the most intense hurricanes to strike major cities— such as a Category 5 storm striking Miami. As a result, expected damage is typically higher than actual damage in most years, although it is less than actual damage in an exceptionally high-cost year. Because expected damage does not depend on the idiosyncratic factors that determine actual hurricane damage, changes in expected damage reflect only changes in underlying conditions.

Unlike *actual* damage, which varies substantially from year to year, *expected* damage is constant for a given set of climatic and demographic conditions (see Box 1). CBO's analysis isolates the effects of long-term trends caused by climate change and coastal development by comparing expected damage at various points in time.

Overview of CBO's Methodology

CBO's analysis involved three steps. First, the agency estimated expected hurricane damage under current conditions—that is, current sea levels and hurricane frequency, as well as current population and per capita income in coastal areas. That estimate served as a "reference case" against which to compare expected damage in future years.

Next, CBO projected how those conditions would change over time. Specifically, for 2025, 2050, and 2075, CBO estimated the likelihood of potential changes of various amounts in each of the four types of underlying conditions—that is, changes in sea levels for affected states, hurricane frequency, population in counties vulnerable to hurricane damage, and per capita income in those counties—that would lead to differences in expected hurricane damage. For each of those conditions, the agency's estimates indicated a range of possible values and the probability that the actual value would fall within various sections of that range.

Finally, CBO determined expected damage under the new conditions for 2025, 2050, and 2075. To do so, the agency produced a distribution of expected damage by simulating expected damage for each of those years 5,000 times. Each simulation used a unique set of projected changes for each of those four underlying conditions—obtained by making random draws (or selections) from the distributions of potential changes in those conditions that CBO estimated. (That technique is often referred to as a Monte Carlo analysis.) The agency assessed changes in expected damage over time by comparing the distributions of such damage in the selected future years to expected damage in the reference case. For each distribution of expected damage in a future year, CBO reports the mean (that is, the average) and the "likely range," which includes the middle two-thirds of the distribution of estimates from the simulations. The accompanying working paper provides a detailed discussion of the data and methodology CBO used to estimate hurricane damage.[3]

Step 1. Estimating Expected Damage on the Basis of Current Conditions: The Reference Case

The first step in estimating changes in future hurricane damage involved creating a reference case of expected damage under current conditions for climate and coastal development. CBO's estimate, which includes the adverse effects of high winds and storm surges on property, contents, and business activity, was obtained using damage functions provided by Risk Management Solutions, or RMS (see Box 2 on page 8).[4] Those damage functions incorporate the probability of landfall at various locations for hurricanes of different categories and indicate expected damage, taking into account the following:

- Existing sea levels in each state,
- The expected number of hurricanes of each category that will make landfall in the United States under current conditions, and
- The current value of residential and nonresidential property exposed to hurricane damage in each state.

In CBO's reference case, wind damage accounts for 64 percent of overall expected damage from hurricanes, and storm surges account for 36 percent; however, the actual percentages can vary widely from storm to storm (see the table in Box 2). Hurricane damage is expected to be greater in some coastal areas than in others. The top three states in terms of expected damage (in descending order) are Florida, Texas, and Louisiana. Those states make up over three-quarters of the total expected annual hurricane damage; Florida alone accounts for more than half of the damage (see Figure 1).

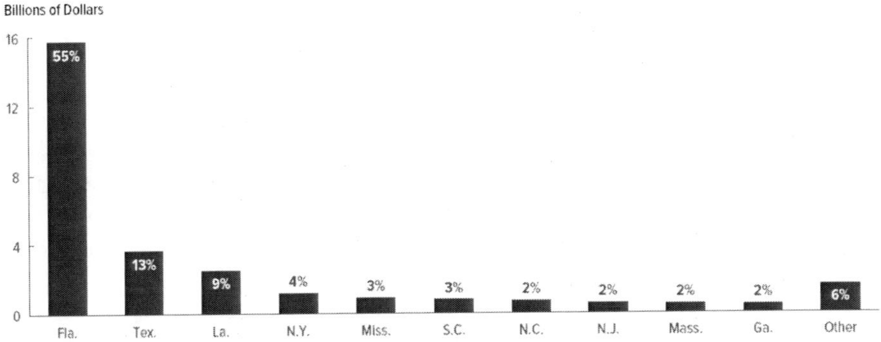

Source: Congressional Budget Office, using data from Risk Management Solutions.
CBO's estimate of expected annual hurricane damage at present—the "reference case"—is based on estimates of the current frequency of hurricanes, current state-specific sea levels, and the current valuation of property exposed to hurricanes.
Percentages indicate each state's share of the total damage in CBO's reference case. Dollar amounts are expressed in 2015 dollars.
"Other" includes the following: Alabama, Connecticut, Delaware, Maine, Maryland, New Hampshire, Pennsylvania, Rhode Island, Vermont, Virginia, Washington, D.C., and West Virginia.

Figure 1. Expected Annual Hurricane Damage in CBO's Reference Case, by State.

Step 2. Estimating Changes in the Four Types of Underlying Conditions That Affect Expected Damage: Inputs to CBO's Analysis

For 2025, 2050, and 2075, CBO estimated the range and likelihood of potential changes in each of the four types of underlying conditions that would lead to changes in expected hurricane damage: sea levels (which vary by state), the frequency of hurricanes, population in counties that are vulnerable to hurricane damage, and per capita income in those same counties.

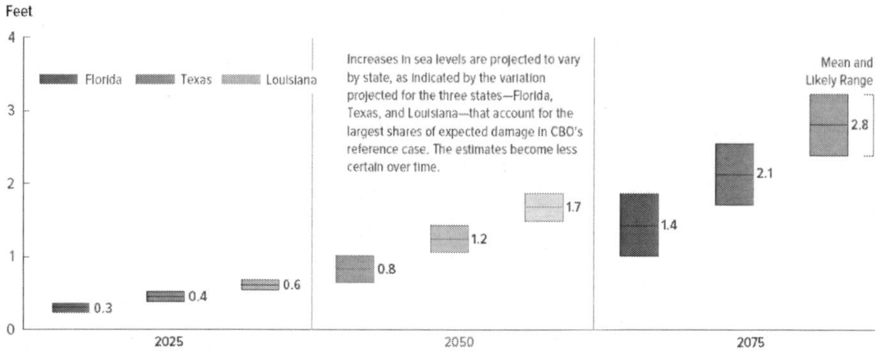

Source: Congressional Budget Office, using data from Risk Management Solutions; Robert E. Kopp and others, "Probabilistic 21st and 22nd Century Sea-Level Projections at a Global Network of Tide-Gauge Sites," *Earth's Future*, vol. 2, no. 8 (August 2014; corrected, October 2014), http://onlinelibrary.wiley.com/doi/10.1002/2014EF000239/full.

CBO's estimate of expected annual hurricane damage at present—the "reference case"—is based on estimates of the current frequency of hurricanes, current state-specific sea levels, and the current valuation of property exposed to hurricanes.

The "likely range" includes the middle two-thirds of the distribution of estimates from CBO's simulations.

Figure 2. Projected Rise of Sea Levels in Florida, Texas, and Louisiana in Selected Future Years.

Effects of Climate Change on Underlying Conditions

Two of the four types of conditions that underlie CBO's analysis—rising sea levels and changes in the frequency of hurricanes—are attributable to climate change.

Changes in Sea Levels

Scientists generally expect sea levels to rise as a result of climate change, but the magnitude of that increase is uncertain. That uncertainty is complicated by the fact that such increases will vary along the East and Gulf coasts. In its analysis, CBO anticipated that future sea levels will change over time and vary by state, consistent—by design—with the range of increases in sea levels that the Intergovernmental Panel on Climate Change (IPCC) predicts will occur globally.[5]

Box 2. Risk Management Solution's Damage Functions

Risk Management Solution's (RMS's) damage functions—which translate hurricane occurrences in the United States into state-specific estimates of expected damage—were determined by simulating tens of thousands of physically realistic hurricane seasons under current conditions, including current sea levels and property exposure in coastal areas. In those simulations, the number of hurricanes of each category (the number of Category 3 storms, for example) making landfall in the United States is consistent with experience during the period from 1900 through 2014. However, the locations of impact are not limited to those of historical storms; in contrast, the simulations account for all the *possible paths* that a given hurricane could follow.

RMS differs from the National Oceanic and Atmospheric Administration (NOAA) in how it categorizes hurricanes. RMS considers their intensity in *advance* of landfall, whereas NOAA—which records historical occurrences—categorizes hurricanes at the *point* of landfall. Because hurricanes generally lose energy as they begin to interact with land—becoming lower-category hurricanes or tropical storms—that difference in categorization means RMS's estimates of hurricane frequencies do not match those that NOAA recorded between 1900 and 2014; RMS's method of categorizing storms results in a greater number of occurrences of all categories of hurricanes.

Because RMS's calculation of expected damage is based on the hurricane's entire wind field and the extent of storm surges as the hurricane moves over land, its estimate of expected damage is not affected by the manner in which it categorizes hurricanes. For example, a hurricane categorized by RMS as a Category 4 storm (based on its wind speed measured at a point in advance of landfall) might have Category 3 wind speeds when it actually makes landfall; in that case, RMS's damage estimate for the storm would be consistent with the Category 3 wind speeds.

According to RMS's simulations, which serve as the basis for the Congressional Budget Office's reference case, Category 1 hurricanes have an expected annual frequency of 0.76, implying that one such hurricane makes landfall, on average, once every 1.3 years (see the table). Although Category 1 storms are the most frequent type of hurricane, they account for less than 8 percent of the total expected hurricane damage of $28 billion in CBO's reference case. (Expected annual damage for storms of a given category is the product of the annual frequency with which such storms occur and the average damage that they create when they do occur.) In contrast, the most intense hurricanes—Category 5 events—account for about 12 percent of total damage even though they only occur (according to RMS's method of measurement), on average, fewer than four times per century. The largest fraction of expected damage comes from Category 4 storms, which, on the basis of RMS's method of categorizing frequency, occur once every four years, on average.

Hurricane Characteristics and Estimated Expected Damage in CBO's Reference Case, by Category of Hurricane

				Expected Annual Damage			
					Percentage		
Category of Hurricane[a]	Wind Speed (Miles per hour)	Expected Frequency per Year[b]	Expected Number of Years Between Storms	Billions of Dollars	From Storm Surges	From Wind	Of Total
1	74–95	0.76	1.3	2	45	55	8
2	96–110	0.34	3.0	3	43	57	12
3	111–129	0.32	3.1	9	45	55	31
4	130–156	0.25	4.1	11	30	70	38
5	157+	0.04	26.3	3	22	78	12
Total or Average		1.70	0.6[c]	28	36	64	100

Source: Congressional Budget Office, using data from Risk Management Solutions (RMS).

CBO's estimate of expected annual hurricane damage at present—the "reference case"—is obtained from RMS and is based on current hurricane frequencies, state-specific sea levels, and the valuation of property exposure by state.

Expected annual damage is expressed in 2015 dollars.

[a] There are five categories of hurricane, which are classified on the basis of their peak wind speed. Category 5 storms are considered the most intense.

[b] The frequency of hurricanes indicates the estimated average number of occurrences in a given year. The frequencies in this table are based on data from RMS, which categorizes hurricanes in advance of landfall. In contrast, the National Oceanic and Atmospheric Administration (NOAA), which records historic occurrences, categorizes hurricanes at the point of landfall. Because hurricanes typically lose energy as they begin to interact with land, RMS's measures of frequency are not comparable to the historic occurrences reported by NOAA. (However, because RMS's calculation of expected damage is based on the hurricane's entire wind field and the extent of storm surges as the hurricane moves over land, its estimate of expected damage is not affected by the manner in which it categorizes hurricanes.)

[c] The expected number of years between hurricanes of any category.

Increases in sea levels—for any given change in the climate—will vary among states for several reasons, including differences in heat content and salinity at various locations.[6] That variation among states can be significant. For example, Florida, which accounts for the largest share of expected damage in the reference case, has the smallest projected average increase in sea level among the 22 states included in CBO's analysis. The average increase for Florida is projected to be 0.8 feet in 2050 and 1.4 feet in 2075 (see Figure 2). By contrast, the average increase for Texas, which accounts for the second-largest share of expected damage, is 1.2 feet in 2050 and 2.1 feet in 2075. The state with the largest projected increase in sea level is Louisiana, which is projected to experience an average rise of 1.7 feet in 2050 and 2.8 feet in 2075.

Louisiana accounts for the third-largest share of expected damage in the reference case.

Changes in the Annual Frequency of Hurricanes

Although scientists find that climate change will affect the conditions that give rise to hurricanes, significant uncertainty surrounds the ultimate effect of climate change on the frequency of hurricanes in the United States. That effect is unclear because of uncertainties about the following:

- Future emissions of greenhouse gases,
- The effects of those emissions on the climatic conditions that affect hurricane formation in the North Atlantic Basin, and
- The relationship between those conditions and the frequency of hurricanes of different categories.

Climate change is expected to alter some conditions in ways that *increase* the probability of hurricanes forming in the North Atlantic and to change other conditions in ways that *decrease* that probability. For example, climate change is expected to increase sea-surface temperatures, which, in turn, increases instability in the atmosphere and creates conditions favorable to hurricane formation.[7] By contrast, climate change is also expected to increase wind shear over the tropical Atlantic, which tends to disrupt the formation of hurricanes. (Wind shear is the difference in the direction and speed of wind over a relatively short distance in the atmosphere.)

For its analysis, CBO used 18 different sets of projections of annual hurricane frequency in the United States. (Each set consisted of a projection of frequency for each of the five categories of hurricanes). Those sets, which were developed by two leading researchers, Thomas Knutson and Kerry

Emanuel, include a wide range of possible outcomes but generally find an increased frequency of major hurricanes (Category 3 and higher; see Figure 3).[8]

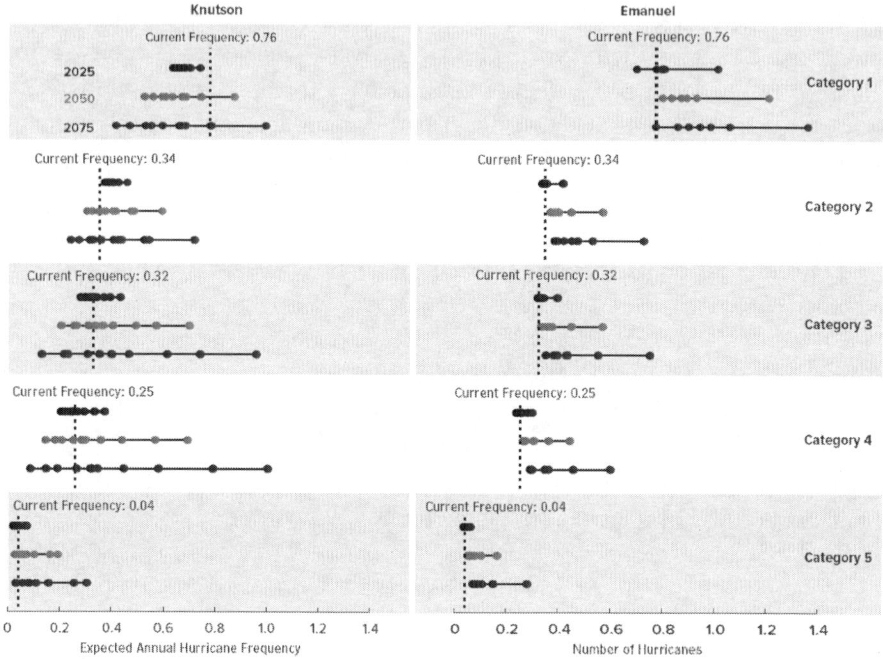

Estimates of hurricane frequency vary by researcher. In general, Knutson more consistently estimates increases in the frequency of Category 4 and 5 hurricanes than in the frequency of Category 1 and 2 hurricanes. By contrast, Emanuel generally estimates increases in the frequency of all categories of hurricanes. Both researchers' projections become less certain over time.

Source: Congressional Budget Office, using information from Kerry A. Emanuel, "Downscaling CMIP5 Climate Models Shows Increased Tropical Cyclone Activity Over the 21st Century," *Proceedings of the National Academy of Sciences*, vol. 110, no. 30 (July 2013), www.pnas.org/content/110/30/12219; Thomas R. Knutson and others, "Dynamical Downscaling Projections of Twenty-First-Century Atlantic Hurricane Activity: CMIP3 and CMIP5 Model-Based Scenarios," *Journal of Climate*, vol. 26, no. 17 (September 2013), http://journals.ametsoc.org/doi/abs/10.1175/JCLI-D-12-00539.1; and additional information from Emanuel and Knutson.

Each circle indicates a projection made by the researcher on the basis of a unique set of factors that influence hurricanes, such as sea surface temperature and wind shear. Those factors were obtained from various atmospheric oceanic general circulation models, with each model projecting outcomes based on a given concentration of greenhouse gases in the atmosphere.

Frequency indicates the estimated average number of occurrences in a given year.

Figure 3. Expected Hurricane Frequency by Researcher, Category, and Year.

Effects of Coastal Development on Underlying Conditions

Continued development along the East and Gulf coasts of the United States is likely to increase hurricane damage simply by increasing the amount of property that is exposed to damage. CBO approximated increases in property exposure by estimating increases in population and per capita income by county.

CBO's model included 777 counties, situated in 22 states, that were found to have some probability of hurricane damage. For each selected year—2025, 2050, and 2075—CBO produced a set of estimates of each county's population and per capita income. Those estimates were based on a combination of the counties' individual historical growth rates and the projected growth rate for the United States as a whole. To account for the uncertainty inherent in those estimates, CBO built a distribution around each estimate, allowing unforeseen circumstances to affect either individual counties or the regions in which they are located (see the working paper that accompanies this report). Those distributions included values above the projected average (resulting from some positive influence, such as the construction of a new manufacturing plant) and values below it (resulting from a negative influence, such as a major oil spill).

Step 3. Constructing Estimates of Expected Damage: The Simulations

CBO simulated expected damage for future selected years thousands of times. Each simulation used the damage functions described above (in the discussion of the reference case) and a unique set of randomly selected projections of the four types of underlying conditions that affect hurricane damage: state-specific sea levels, the frequency of hurricanes, county-specific populations, and county-specific per capita income.

CBO used the projections of sea levels and the frequency of hurricanes, along with the damage functions provided by RMS, to assess a range of possible effects of climate change on expected damage in each state. CBO adjusted the resulting "climate change only" damage estimate for each state on the basis of the agency's projections of population increases in the state's counties and per capita income. CBO's adjustment accounted for the relative vulnerability of counties within a given state to damage from wind and storm surges (giving more weight to growth in population and per capita income in relatively vulnerable counties).

To determine a set of stylized "medium response" estimates, the agency relied on its own judgment and on the limited literature that is available on the response of damage to changes in population and per capita income in coastal areas:

- *Responses to Changes in Population*. CBO estimated that a 10 percent increase in population would trigger a relatively small (2.5 percent) increase in damage from wind and a somewhat larger (5 percent) increase in damage from storm surges. The responses reflect the fact that greater density would generate some protection from wind damage and, in the case of taller structures, from storm surges.
- *Responses to Changes in per Capita Income.* CBO estimated that a 10 percent increase in per capita income would trigger a proportionate (10 percent) increase in damage from wind and a less than proportionate (7.5 percent) increase in damage from storm surges. Those responses reflect the judgment that increases in income might result in upgrades to infrastructure—such as the construction of seawalls or the raising of buildings—that would better guard against storm surges than wind.

CBO also projected expected damage under alternative assessments of the effect of growth in the population and per capita income on the growth in damage (described below).

RESULTS: ESTIMATES OF EXPECTED HURRICANE DAMAGE FOR SELECTED FUTURE YEARS

The combined effects of climate change and coastal development will cause hurricane damage to increase in the future, CBO estimates—both in dollar terms (expressed as a share of GDP, which provides a measure of the nation's ability to pay for the damage), and in terms of the percentage of the U.S. population living in counties where expected damage is particularly burdensome.

Damage as a Share of GDP

The growth in expected hurricane damage is expected to exceed the growth in the nation's ability to pay for such damage. The U.S. economy, and thus the nation's ability to pay for hurricane damage, is projected to grow over time: CBO estimates that GDP will be nearly four times larger in 2075 than it is today. However, CBO also estimates that the combined forces of climate change and coastal development will cause expected hurricane damage to grow more quickly than the size of the economy.

The two measures would grow proportionately through 2025: Mean expected damage in 2025 is projected to be 0.16 percent of GDP in that year, essentially the same share that is projected in CBO's reference case (0.16 percent of GDP—or about $28 billion—today). But mean expected damage rises to 0.19 percent in 2050 and to 0.22 percent in 2075, relative to projected GDP for those years (see Figure 4). Applied to today's economy, those percentages would indicate annual expected damage of about $34 billion and $39 billion, respectively.

Uncertainty about those estimates grows over time. For example, the likely range of expected damage is one-third wider in 2075 (when the difference between the upper and lower ends of the likely range amounts to 0.16 percent of GDP) than in 2050 (when the difference amounts to 0.12 percent of GDP).[9] Based on today's economy, the middle two-thirds of CBO's estimates in 2075 lie between $27 billion and $56 billion.

The Relative Contribution of Climate Change and Coastal Development to Increasing Damage

Climate change and coastal development will occur simultaneously, and each factor will compound the increase in expected damage caused by the other. For example, rising sea levels—and the resulting increase in expected damage from storm surges—compound the increase in expected damage resulting from expanding state populations. As a result, the combined effects of climate change and coastal development increase expected damage by a greater amount than the sum of the increase in expected damage that each would bring about on its own.[10]

Considering those combined effects, CBO attributes roughly 45 percent of the increase in expected damage in 2075—that is, the rise from 0.16 percent of GDP in the reference case to 0.22 percent of GDP in 2075—to climate change and 55 percent to coastal development.[11]

Sensitivity of Results to Estimates of Hurricane Frequency by Different Researchers

The results discussed above are based on CBO's combination of estimates of hurricane frequency developed by Knutson and Emanuel. CBO repeated the analysis using predictions of hurricane frequency made by each researcher. That sensitivity analysis indicated that average expected damage was not sensitive to the choice of researcher, varying by only 0.01 percent of GDP (see Table 1).

In contrast, the width of the likely range varied significantly depending on which researcher's estimates were used. Relative to the likely range in 2075 that was obtained when both researchers' predictions were used (from 0.15 to 0.31 percent of GDP), the range was much wider (from 0.10 to 0.34 percent of GDP) when only Knutson's predictions were used but much narrower (from 0.17 to 0.26 percent of GDP) when only Emanuel's predictions were used.

Sensitivity of Results to Alternative Estimates of How Much Changes in Population and per Capita Income Affect Estimates of Damage

CBO's estimates of expected damage are sensitive to assessments of how much increases in population and per capita income will increase hurricane damage. Given the importance of those effects, CBO examined the extent to which its estimate of expected damage in 2075 would change if hurricane damage was more, or less, sensitive to changes in population and per capita income. Specifically, CBO calculated a distribution of damage in 2075 under two alternative scenarios:

- *Higher-response case.* CBO assumed that damage associated with a 10 percent increase in population or per capita income is 2.5 percentage points greater than in the medium-response case described above. For example, rather than a 10 percent increase in per capita income resulting in a 10 percent increase in damage from wind, as was estimated in the medium-response case, a 10 percent increase in per capita income results in a 12.5 percent increase in damage in the higher-response case (see Table 2).[12] That change makes expected damage more sensitive to increases in population and per capita income, implying that communities would develop in a way that would not limit expected damage (for example, if increases in population did not increase housing density enough to limit wind damage) and that communities and individuals would make fewer adjustments (such as constructing storm walls) to limit potential damage from hurricanes.

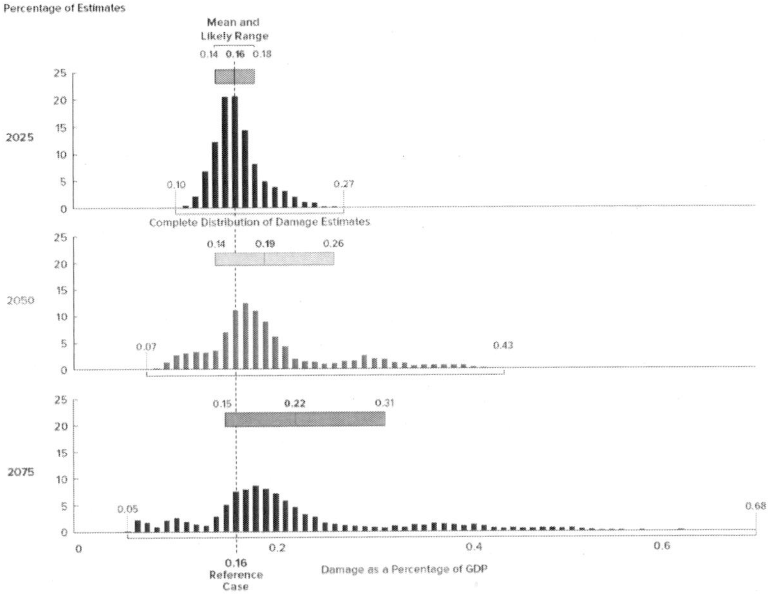

CBO estimates that expected hurricane damage, measured as a percentage of GDP, will be nearly 40 percent higher (at 0.22 percent) in 2075 than under current conditions (0.16 percent). The uncertainty surrounding estimates of expected damage grows substantially over time.

Source: Congressional Budget Office, using data from Risk Management Solutions; Kerry A. Emanuel, "Downscaling CMIP5 Climate Models Shows Increased Tropical Cyclone Activity Over the 21st Century," *Proceedings of the National Academy of Sciences*, vol. 110, no. 30 (July 2013), www.pnas.org/ content/110/30/12219; Thomas R. Knutson and others, "Dynamical Downscaling Projections of Twenty-First-Century Atlantic Hurricane Activity: CMIP3 and CMIP5 Model-Based Scenarios," *Journal of Climate*, vol. 26, no. 17 (September 2013), http://journals.ametsoc.org/ doi/abs/10.1175/JCLI-D-12-00539.1; and additional information provided by Emanuel and Knutson.

CBO's estimate of expected annual hurricane damage at present—the "reference case"—is based on estimates of the current frequency of hurricanes, current state-specific sea levels, and the current valuation of property exposed to hurricanes.

CBO's estimates of the distribution of expected hurricane damage in selected years are based on the results of 5,000 simulations, with each simulation using a unique set of draws (random selections) for underlying conditions that determine expected damage.

The "likely range" includes the middle two-thirds of the distribution of estimates from CBO's simulations. GDP = gross domestic product.

Figure 4. Estimates of Expected Hurricane Damage as a Percentage of GDP in Selected Future Years.

Table 1. Estimates of Expected Damage in 2075, Based on Projections of Hurricane Frequency by Two Researchers

Percentage of GDP				
			Likely Range	
Researcher	Mean	Low End	High End	Width
Knutson	0.21	0.10	0.34	0.24
Emanuel	0.22	0.17	0.26	0.09
Both	0.22	0.15	0.31	0.16

Source: Congressional Budget Office, using data from Risk Management Solutions; Kerry A. Emanuel, "Downscaling CMIP5 Climate Models Shows Increased Tropical Cyclone Activity Over the 21st Century," *Proceedings of the National Academy of Sciences*, vol. 110, no. 30 (July 2013), www.pnas.org/content/ 110/30/12219; and Thomas R. Knutson and others, "Dynamical Downscaling Projections of Twenty-First-Century Atlantic Hurricane Activity: CMIP3 and CMIP5 Model-Based Scenarios," *Journal of Climate*, vol. 26, no. 17 (September 2013), http://journals.ametsoc.org/ doi/abs/10.1175/JCLI-D-12-00539.1.

The "likely range" includes the middle two-thirds of the distribution of estimates from CBO's simulations.

CBO's main results, which are presented in Figures 4 and 5, combine the predictions of both Knutson and Emanuel.

GDP = gross domestic product.

- *Lower-response case*. CBO assumed that damage associated with a 10 percent increase in population or per capita income is 2.5 percentage points less than in the medium-response case. That change makes expected damage less sensitive to increases in population and per capita income, implying, for example, that increases in population would lead to denser housing or that increases in per capita income would cause individuals and businesses to invest in more hurricane-resistant houses and buildings.

CBO found that the mean estimate of expected damage for 2075 is roughly 20 percent higher in the higher-response case (0.26 percent of GDP) and roughly 20 percent lower in the lower-response case (0.17 percent of GDP) than in the medium-response case (0.22 percent of GDP).

Percentage of the U.S. Population Facing Substantial Expected Damage

The expected damage from hurricanes will constitute a larger share of per capita income for people living along the East and Gulf coasts than for the average person in the United States (approximated by damage as a share of GDP). Currently, roughly 1.2 million people—0.4 percent of the U.S. population—live in counties facing substantial expected damage. For the purposes of this analysis, "substantial expected damage" is defined as expected per capita damage that is greater than 5 percent of the county's average per capita income. CBO projects that by 2050, 5.8 million people—1.4 percent of the population—will face expected damage that exceeds the 5 percent threshold and that by 2075 those amounts will rise to 10 million people, or 2.1 percent of the population (see Figure 5 on page 16). The percentage of the U.S. population likely to exceed the 5 percent threshold in 2075 ranges from 0.3 percent (or 1.6 million people) to 5.2 percent (or about 25 million people).

The low end of the likely range includes the people in 13 counties, while the high end includes the people in 68 counties (roughly five times as many counties). Because CBO estimates that damage grows more slowly than population, the agency's estimates of per capita damage in 2075 tend to be higher in states with smaller populations than in states with larger populations. Consequently, the 13 counties exceeding the 5 percent threshold at the low end of the likely range (corresponding to lower damage estimates) have an average population that is less than one-third of the average population for the 68 counties exceeding the threshold at the high end of the likely range (corresponding to higher estimates of damage). This result leads to the very wide likely range that CBO estimates.

Budgetary Implications of Hurricane Damage

Hurricane damage can affect the federal budget in a variety of ways. When a disaster overwhelms the capacity of state and local resources to provide relief, the governor of a state or a tribal leader may request federal assistance under the Stafford Disaster Relief and Emergency Assistance Act (the Stafford Act). In response, the President may declare such an event to be a major disaster, which authorizes certain federal agencies to provide various types of financial and technical assistance to state, local, and tribal governments, as well as to individuals and businesses in affected areas.

Table 2. The Extent to Which Changes in Population and per Capita Income Affect Estimates of Expected Damage in 2075

	Medium Response[a]	Higher Response	Lower Response
Change in Wind Damage Given a 10 Percent Change in . . .			
Per capita income	10.0	12.5	7.5
Population	2.5	5.0	0
Change in Storm-Surge Damage Given a 10 Percent Change in . . .			
Per capita income	7.5	10.0	5.0
Population	5.0	7.5	2.5
Expected Damage (Percentage of GDP)			
Mean	0.22	0.26	0.17
Likely range[b]			
Low end	0.15	0.17	0.11
High end	0.31	0.37	0.25

Source: Congressional Budget Office, using data from Risk Management Solutions; Kerry A. Emanuel, "Downscaling CMIP5 Climate Models Shows Increased Tropical Cyclone Activity Over the 21st Century," *Proceedings of the National Academy of Sciences*, vol. 110, no. 30 (July 2013), www.pnas.org/content/110/30/12219; and Thomas R. Knutson and others, "Dynamical Downscaling Projections of Twenty-First-Century Atlantic Hurricane Activity: CMIP3 and CMIP5 Model-Based Scenarios," *Journal of Climate*, vol. 26, no. 17 (September 2013), http://journals.ametsoc.org/doi/abs/10.1175/JCLI-D-12-00539.1.

GDP = gross domestic product.

[a] The changes in wind damage and storm-surge damage that constitute the "medium response" underlie CBO's estimates of hurricane damage and the affected population, which are provided in Figures 4 and 5.

[b] The "likely range" includes the middle two-thirds of the distribution of estimates from CBO's simulations.

For the most part, decisions about whether to provide federal funding for disaster relief in the aftermath of a hurricane, about amounts of funding for specific agencies and programs, and about what conditions should be placed on the spending are not mandated by law. Rather, such funding is largely discretionary and is the outcome of choices made by policymakers on a case-by-case basis. The disaster-related accounts, such as FEMA's Disaster Relief Fund, generally receive annual appropriations sufficient for noncatastrophic events, but lawmakers typically rely on supplemental appropriations to provide sufficient funds to respond to major hurricanes.

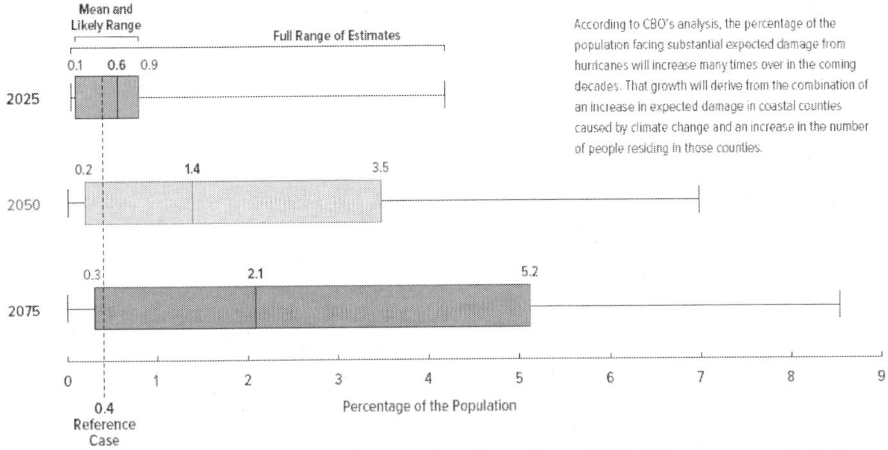

Source: Congressional Budget Office, using data from Risk Management Solutions; Kerry A. Emanuel, "Downscaling CMIP5 Climate Models Shows Increased Tropical Cyclone Activity Over the 21st Century," *Proceedings of the National Academy of Sciences*, vol. 110, no. 30 (July 2013), www.pnas.org/content/110/30/12219; and Thomas R. Knutson and others, "Dynamical Downscaling Projections of Twenty-First-Century Atlantic Hurricane Activity: CMIP3 and CMIP5 Model-Based Scenarios," *Journal of Climate*, vol. 26, no. 17 (September 2013), http://journals.ametsoc.org/doi/abs/10.1175/JCLI-D-12-00539.1.

People exposed to substantial hurricane damage are defined as those living in counties in which per capita expected damage is greater than 5 percent of the county's per capita income.

CBO's estimate of expected annual hurricane damage at present—the "reference case"—is based on estimates of the current frequency of hurricanes, current state-specific sea levels, and the current valuation of property exposed to hurricanes.

CBO's estimates of the distribution of expected hurricane damage in selected years are based on the results of 5,000 simulations, with each simulation using a unique set of draws (random selections) for four underlying conditions that determine expected damage.

The "likely range" includes the middle two-thirds of the distribution of estimates from CBO's simulations.

Figure 5. Percentage of the U.S. Population Living in Counties With Substantial Expected Hurricane Damage in Selected Future Years.

There is one exception to that general rule. Unlike spending from disaster-relief programs that are funded through discretionary appropriations, claim payments made through the National Flood Insurance Program (NFIP) are a form of mandatory spending—that is, the government is required by statute to make those payments. To the extent that those payments are not fully covered by insurance premiums, they represent costs to the federal government (in addition to the discretionary appropriations).

Hurricane damage can also have indirect effects on the federal budget through changes that it causes in the overall amount of economic activity (GDP) and the use of income support programs and special tax provisions. Those indirect effects were not quantified in CBO's analysis, but as discussed on pages 20 through 21, they are likely to be relatively small.

Since Hurricane Katrina, federal spending for the storms that were included in CBO's analysis—specifically, those that occurred between August 2005 and December 2015 and resulted in at least $1 billion in damage—has exceeded 60 percent of total hurricane damage. CBO estimates that climate change and coastal development will result in an increase in expected hurricane damage measured as a share of GDP. That increase, in turn, will lead to an increase in federal spending as a share of GDP unless disaster relief, measured as a share of damage, falls enough to compensate.

Discretionary Spending

Discretionary spending aimed at helping individuals, businesses, and communities address hurricane damage has totaled $209 billion since 2000 (see Table 3). More than half of that amount was spent responding to Hurricane Katrina, and one-quarter was spent in the aftermath of Hurricane Sandy.

Many federal agencies help administer and coordinate assistance to communities affected by hurricane damage.[13] However, over the past 15 years, roughly three-quarters of federal funding for such disaster assistance has been spent through programs administered by three agencies: the Federal Emergency Management Agency, the Department of Housing and Urban Development (HUD), and the Army Corps of Engineers (see Figure 6).

Federal Emergency Management Agency: Disaster Relief Fund

FEMA administers the Disaster Relief Fund, which accounted for about 45 percent of all discretionary spending on hurricanes over the past 15 years.

The DRF is used to pay for current emergencies and ongoing recovery projects from past disasters.[14] Since 2000, it has provided $95 billion for hurricane relief, mostly funded by supplemental appropriations. Three programs account for the bulk of spending from the DRF (see Figure 7).

Table 3. Total Federal Spending and Total Economic Damage for Selected Hurricanes, 2000 to 2015

Hurricane	Date	Total Federal Spending in Response to Hurricanes (Billions of dollars)			Percentage of Total Federal Spending in Response to Hurricanes	Total Economic Damage (Billions of dollars)	Percentage of Total Economic Damage
		Discretionary	Mandatory	Total			
Lili	Oct-02	0.1	0	0.1	0.1	1.3	0.3
Isabel	Sep-03	1.1	0.1	1.2	0.6	7.7	1.9
Charley	Aug-04	2.0	0	2.1	1.0	20.9	5.2
Frances	Sep-04	2.8	0	2.8	1.3	12.3	3.0
Ivan	Sep-04	3.8	0.1	3.9	1.8	25.8	6.4
Jeanne	Sep-04	3.3	0	3.3	1.6	9.8	2.4
Dennis	Jul-05	0.3	0	0.3	0.2	3.5	0.9
Katrina	Aug-05	107.9	2.3	110.2	51.3	147.4	36.5
Rita	Sep-05	8.6	0.1	8.7	4.0	22.4	5.5
Wilma	Oct-05	6.1	0.1	6.2	2.9	22.4	5.5
Dolly	Jul-08	0.2	0	0.3	0.1	1.1	0.3
Gustav	Sep-08	4.0	0	4.0	1.9	6.7	1.7
Ike	Sep-08	12.0	0.4	12.3	5.7	33.7	8.4
Irene	Aug-11	3.8	0.5	4.3	2.0	15.5	3.8
Isaac	Aug-12	1.2	0.0	1.3	0.6	3.2	0.8
Sandy	Oct-12	51.2	2.6	53.9	25.1	69.9	17.3
Total		208.6	6.3	214.9	100.0	403.6	100.0

Source: Congressional Budget Office, using information from the sources listed in Appendix B.

Estimates of total economic damage are from the National Oceanic and Atmospheric Administration, National Hurricane Center, "Tropical Cyclone Reports" (June 5, 2015), www.nhc.noaa.gov/data/tcr/index.php?season=2014&basin=atl.

Dollar amounts are expressed in 2015 dollars.

Includes all hurricanes resulting in at least $1 billion in damage between 2000 and 2015. (Not all years included such hurricanes.)

Public Assistance Program

The largest of the three programs, the Public Assistance Program, helps communities cover the costs of repairing or replacing public buildings, infrastructure, and utilities, as well as emergency measures, such as removing debris, establishing temporary shelters, and providing emergency power. The federal government covers at least 75 percent of the cost of activities that are eligible for such assistance, but that share can be higher if certain per capita damage thresholds are met. From fiscal years 2000 to 2013, support provided through the Public Assistance Program accounted for 47 percent of total spending from the DRF.[15]

Individual Assistance Program

This program accounted for 25 percent of spending from the DRF between fiscal years 2000 and 2013. Individual assistance includes temporary housing for displaced people, grants for the repair or replacement of a home and other damaged property, grants for medical treatment, cleaning, moving, or other needs not covered by insurance, and disaster-related unemployment insurance.

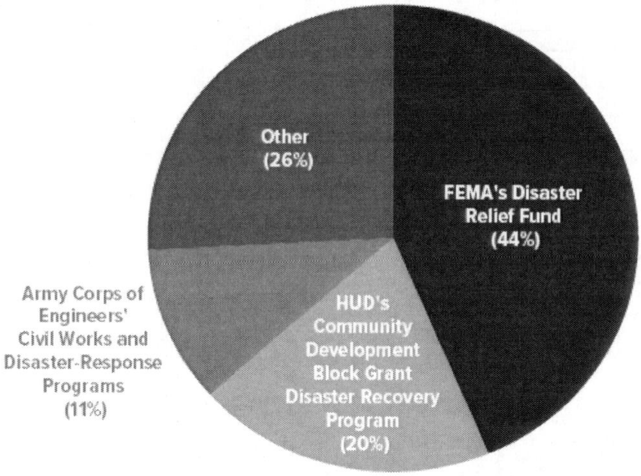

Source: Congressional Budget Office, using information from the sources listed in Appendix B.

"Other" includes the Department of Transportation, Department of Education, Department of Defense (not including the Army Corps of Engineers), and Small Business Administration.

Includes all hurricanes resulting in at least $1 billion in damage between 2000 and 2015. (Not all years included such hurricanes.)

FEMA = Federal Emergency Management Agency; HUD = Department of Housing and Urban Development.

Figure 6. Share of Discretionary Federal Spending on Selected Hurricanes, by Agency, Fiscal Years 2000 to 2015.

Hazard Mitigation Grant Program

This program, which accounted for 6 percent of spending from the DRF between fiscal years 2000 and 2013, supports mitigation measures that are designed to prevent or reduce the loss of life or property damage from a future disaster.[16] Mitigation projects are categorized as structural and nonstructural. Structural activities include retrofitting or strengthening facilities so that they will be more resistant to future damage, elevating structures to reduce flooding, and implementing building codes that are designed to increase structural resiliency. Nonstructural activities include community planning initiatives, creating disaster mitigation and flood plans, and implementing disaster warning systems.

Department of Housing and Urban Development: Community Development Block Grant Disaster Recovery Program

Following major disasters, the Congress has often provided large supplemental appropriations for HUD to fund disaster recovery activities through its Community Development Block Grant Disaster Recovery Program. Such grants have accounted for 20 percent of discretionary spending related to hurricane damage since 2000, and reliance on them has increased over time (particularly since Hurricane Katrina). State and local governments have considerably more leeway about how they may use funds provided by the block grants than is the case with other forms of assistance. Those grants are generally used for activities that fall into three categories.[17]

- *Short-Term Disaster Relief.* Such support meets immediate needs following a disaster. Eligible activities include debris removal not covered by FEMA, relocation payments for displaced individuals and businesses, provision of security patrols in damaged areas, and restoration of essential services, such as water, sewer, power, and telecommunications.
- *Mitigation.* Such activities are designed to anticipate and reduce the amount of damage that a future disaster might cause. They include both structural measures (such as strengthening houses and other buildings or implementing flood-control measures) and nonstructural measures (such as funding training exercises, preparedness plans, and buyouts of real estate that is prone to flooding).
- *Long-Term Recovery.* Such activities are designed to rebuild and economically revitalize affected areas. This type of assistance includes grants to homeowners to repair or replace residences, grants

and loans to businesses that have suffered economic damage, and grants to local governments to fund infrastructure improvements.

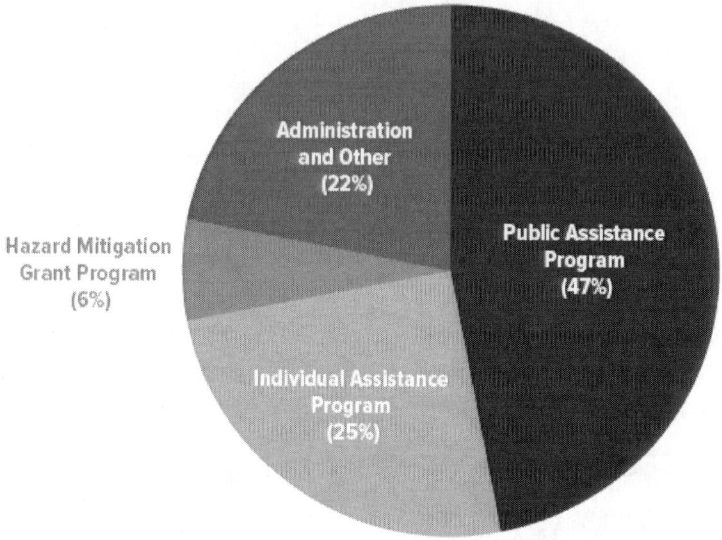

Source: Congressional Budget Office, using data from the Federal Emergency Management Agency.

The Public Assistance Program helps communities cover the costs of repairing or replacing public buildings and infrastructure, restoring utilities, establishing emergency shelters, and so on. The Individual Assistance Program provides temporary housing for displaced people, grants for medical treatment and for the repair of damaged property, and other assistance. The Hazard Mitigation Grant Program includes measures that are designed to prevent or reduce the loss of life or property damage that could result from a future disaster, such as elevating structures to reduce damage from flooding and implementing warning systems. "Administration and Other" includes FEMA's administrative costs, technical assistance contracts, and mission assignments.

FEMA = Federal Emergency Management Agency.

Figure 7. Share of Spending From FEMA's Disaster Relief Fund, by Type of Activity, Fiscal Years 2000 to 2013.

Army Corps of Engineers: Civil Works and Disaster Response

The Army Corps of Engineers administers civil works programs and disaster-response programs that have accounted for 11 percent of discretionary spending related to hurricane damage since 2000. To prevent damage from hurricanes, the Corps constructs and manages an array of flood-control and infrastructure projects, including dams, levees, and seawalls, as well as natural barriers such as sand dunes. The Corps also assists with disaster-response activities, such as debris removal, demolition, commodity distribution, temporary housing, emergency power, and support for urban search and rescue.[18] Those activities are carried out using a variety of contracts awarded in advance that can be quickly activated, usually at FEMA's request.

In recent years, major hurricanes have led the Congress to provide large supplemental appropriations to the Corps to repair infrastructure and control flooding, including over $17 billion to rebuild and strengthen levees and floodwalls damaged by the 2005 hurricanes (primarily Hurricane Katrina) and $5.2 billion to repair damage caused by Hurricane Sandy in 2012.

Other Federal Agencies

Depending on the nature of damage caused by a hurricane, other federal agencies have received notably large supplemental appropriations. For example, in response to the three hurricanes that occurred in 2005 (Katrina, Rita, and Wilma), the Congress provided these amounts (measured in 2015 dollars):

- $9.5 billion to the Department of Defense (not including the Army Corps of Engineers) to fund the repair of military facilities, equipment, and ships, as well as the costs of deployments for relief operations;
- $4.5 billion to the Department of Transportation, primarily for the repair and reconstruction of roads that make up the federal-aid highway system and other critical transportation infrastructure;
- $1.9 billion to the Small Business Administration (SBA) to cover the subsidy costs of low-interest, longterm loans for individuals and businesses to repair and replace homes and property.[19] The SBA used that appropriation to disburse a total of $6.7 billion in disaster loans for the 2005 hurricanes; and
- $2.1 billion to the Department of Education to fund reconstruction of local schools and provide aid to displaced students and teachers.

Supplemental appropriations to the Department of Transportation were particularly large in response to Hurricane Sandy, totaling over $13 billion.

Mandatory Spending: The National Flood Insurance Program

The NFIP was created in 1968 to provide property owners flood insurance, which was virtually unavailable at that time because it would have been prohibitively costly for the private market to provide. The NFIP is the main source of such insurance in the United States.

Eligibility for flood insurance coverage under the program is based on flood-hazard maps developed by FEMA and requires that participating communities comply with building-code requirements and floodplain-management practices set by FEMA. Property owners in participating communities may purchase NFIP policies through private insurance companies.

Because NFIP claim payments are not fully covered by insurance premiums, the gap between payments and premiums represents a cost to the federal government. About one-fifth of NFIP policies—generally those covering older structures in areas at high risk of flooding—are explicitly subsidized by the federal government. On the basis of information provided by NFIP actuaries, CBO estimates that federal subsidies have, on average, covered 50 percent of the premiums for those explicitly subsidized policies, resulting in a total subsidy cost of $6.3 billion for hurricanes occurring since 2000. As discussed below, that sum underestimates federal spending to the extent that FEMA's estimates of flood risks—and the corresponding premiums (including premiums for the four-fifths of policies that are not explicitly subsidized)— are too low.

Indirect Federal Budgetary Effects Not Included in This Analysis

Hurricanes can affect the federal budget in ways beyond those included in this analysis. Major hurricanes can temporarily slow the growth of GDP, for instance, resulting in reductions in revenues. Moreover, the hardship caused by hurricanes can increase the number of people who qualify for means-tested programs and for reductions in taxes.

Effects on Revenues Caused by Changes in GDP

Major hurricanes can have effects on GDP, as well as associated adverse effects on tax revenues. The immediate impact on GDP would be negative. For example, in September 2005, CBO estimated that GDP would grow about one-half of one percent more slowly in the second half of 2005 as a result of Hurricane Katrina, which struck in August of that year.[20] For small changes in GDP, CBO finds that federal tax revenues respond roughly proportionately. Thus, federal revenues that were ultimately collected for taxes owed during that period were probably also about one-half of one percent lower.

The initial negative effect of Katrina on economic output stemmed from lost production in affected regions and from temporary spikes in energy costs that resulted from damage to energy infrastructure. The effects of other hurricanes on GDP, and thus on federal revenues, vary depending on both the magnitude of the damage and the type of damage—the share of damage incurred by businesses and production facilities as opposed to houses, for example.

Following the immediate negative impact, the effects on GDP would depend on the hurricane's impact on investment (spending by businesses to rebuild damaged plants and replace damaged equipment, as well as spending by individuals to repair or replace damaged housing, for example); spending by consumers on durable goods (to replace damaged cars, for instance); and the government's spending for goods and services, and households' consumption expenditures. Rebuilding efforts could even cause growth in GDP to be temporarily higher than it would have been in the absence of the hurricane; however, it is unlikely that the loss in revenues caused by the initial drop in GDP would be fully recovered.

In general, CBO concludes that, over the long run, a single hurricane would have no significant impact on the nation's GDP. That outcome reflects the fact that, in the long run, GDP is determined by the size of the capital stock (which is determined by national saving and capital inflows, neither of which is likely to be permanently affected by the hurricane) and by the labor supply and technological progress (which also are not likely to be permanently affected). In contrast, if climate change were to increase the frequency of major hurricanes, the economy might not fully recover from one catastrophic storm before it was hit by another. Consequently, negative effects on economic output could persist for longer periods of time.[21]

Other Effects on Federal Receipts

In addition to the loss in tax revenues associated with any reduction in the growth of GDP, a major hurricane can delay tax payments collected from households and businesses in affected areas and cause increases in tax deductions associated with personal property losses. Lawmakers also have occasionally enacted legislation providing more extensive assistance on a temporary basis. For example, following Hurricane Katrina, they enacted the Katrina Emergency Tax Relief Act of 2005, which allowed individuals to deduct more personal property losses from their taxable income and gave them more time to replace damaged property without being assessed income taxes on the insurance proceeds. It also allowed businesses and individuals to deduct more charitable donations from taxable income.

Furthermore, a hurricane that affects oil-producing facilities, as was the case with Hurricane Katrina, could cause a reduction in royalty payments made by firms to the federal government for oil and gas leases.[22] In September 2005, CBO estimated that Katrina reduced such payments by a few hundred million dollars, which was more than offset by the roughly $700 million that the government received by selling 11 million barrels of oil from the Strategic Petroleum Reserve in response.[23]

Effects on Spending from Means-Tested Programs

The hardship and loss of income caused by hurricanes can cause people to become eligible for benefits they would not normally collect. For example, people could qualify for support through the Supplemental Nutrition Assistance Program (a program designed to provide nutritional assistance to low-income individuals and families), Medicaid, and unemployment insurance.[24]

Federal Spending Relative to Hurricane Damage

It is difficult to identify how much spending is attributable to a specific hurricane. Emergency supplemental funds, for example, have often been appropriated as part of larger bills that contain funds for multiple disasters (for instance, Hurricanes Katrina, Rita, and Wilma in 2005). To estimate spending for each major hurricane, CBO relied on information from numerous sources, including spending data and program documents from federal agencies, the text of enacted legislation and accompanying reports, and analyses of appropriations and spending produced by the Congressional Research Service. (The sources that CBO used to estimate federal spending are listed in Appendix B.) Since 2000, federal spending on relief from hurricane damage has been dominated by two storms: Hurricane Katrina accounted for over 50 percent of all federal spending between 2000 and 2015, and Hurricane Sandy accounted for 25 percent (see Table 3 on page 17).

Although federal spending as a percentage of hurricane damage varies widely from storm to storm—ranging from a low of 9 percent for Hurricane Lili (which occurred in 2002) to a high of nearly 80 percent for Hurricane Sandy—that percentage has tended to be larger since Hurricane Katrina (see Figure 8). For all of the hurricanes for which CBO had data, federal spending as a percentage of damage rose from 17 percent before Katrina struck to 62 percent from Katrina forward.[25]

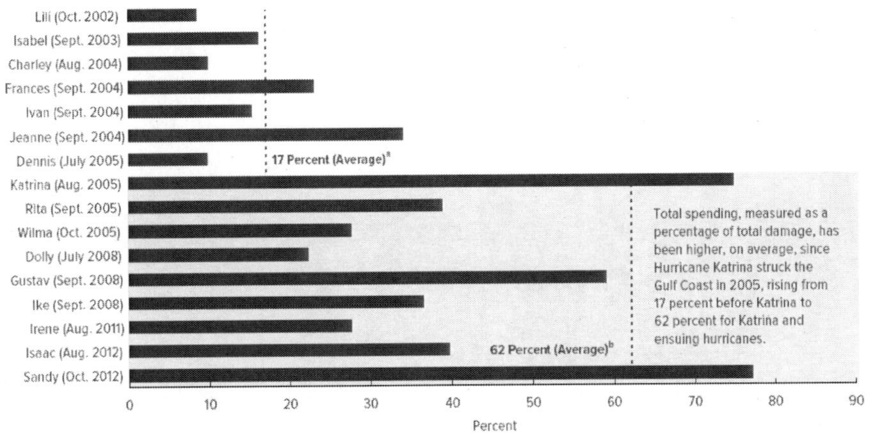

Source: Congressional Budget Office, using information from the sources listed in Appendix B.

The hurricanes included here were ones for which CBO was able to obtain detailed data on federal spending.

Includes all hurricanes resulting in at least $1 billion in damage between 2000 and 2015. (Not all years included such hurricanes.)

[a] Total federal spending as a percentage of total economic damage for hurricanes Lili through Dennis.

[b] Total federal spending as a percentage of total economic damage for hurricanes Katrina through Sandy.

Figure 8. Total Federal Spending as a Percentage of Total Economic Damage for Selected Hurricanes, 2000 to 2015.

FUTURE FEDERAL SPENDING IN RESPONSE TO HURRICANE DAMAGE

If federal spending remained at roughly 60 percent of hurricane damage—a prospect referred to in this report as a historical cost scenario—it would total

0.13 percent of GDP in 2075 (equivalent to about $24 billion in today's economy), compared with 0.10 percent of GDP (or $18 billion) projected under current conditions. That estimate is based on CBO's assessment of *average* expected damage in 2075. Taking into account the likely range of expected damage in 2075, federal spending under the historical cost scenario would range from 0.09 percent to 0.19 percent of GDP, which would amount to between $16 billion and $34 billion in today's economy (see Figure 9).[26]

Because federal spending on hurricane disaster relief is largely discretionary and mostly the result of supplemental appropriations, it could be higher or lower than the 60 percent of hurricane damage that the historical cost scenario implies. Consequently, CBO considered the implications of two alternative measures for federal spending as a percentage of future hurricane damage:

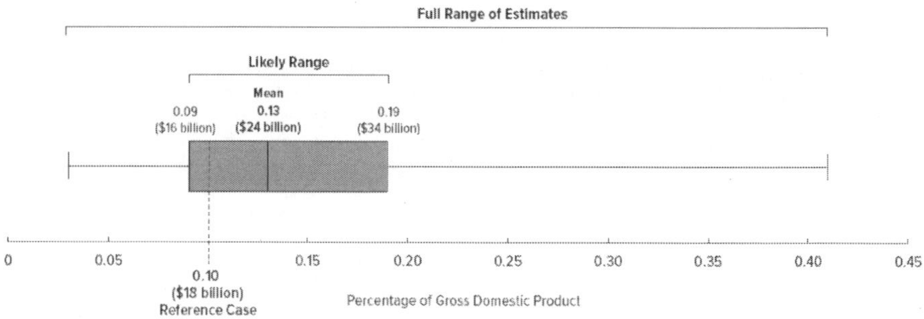

Source: Congressional Budget Office.

Dollar amounts in parentheses are based on today's economy.

In the historical cost scenario, federal spending is estimated to be 60 percent of expected hurricane damage. CBO's estimate of expected annual hurricane damage at present—the "reference case"—is based on estimates of the current frequency of hurricanes, current state-specific sea levels, and the current valuation of property exposed to hurricanes. CBO's estimates of the distribution of expected hurricane damage in 2075 are based on the results of 5,000 simulations, with each simulation using a unique set of draws (random selections) for the underlying conditions that determine expected damage.

The "likely range" includes the middle two-thirds of the distribution of estimates from CBO's simulations.

Figure 9. Estimates of Federal Spending on Hurricane Damage in 2075, Based on the Historical Cost Scenario.

40 percent and 80 percent. Based on the likely range of hurricane damage estimated in the previous section and those two alternative measures, federal

spending resulting from hurricane damage in 2075 could range from 0.06 percent to 0.25 percent of GDP, which would equal $11 billion to $45 billion in today's economy (see Table 4).

APPROACHES TO REDUCING FEDERAL SPENDING ON HURRICANE DAMAGE

In considering how policymakers might seek to reduce future federal spending on hurricane damage, CBO examined several diverse options: limiting greenhouse gas emissions; shifting a greater share of the cost of hurricane damage onto households, businesses, and state and local governments, thereby reducing coastal development; and investing in structural changes, such as elevating buildings and roads, that would cause vulnerable coastal areas to suffer less damage when hurricanes occur.

Limit Greenhouse Gas Emissions

A coordinated effort to significantly reduce global emissions could lessen the potential for increased hurricane damage in the United States between now and 2075; but, the extent of the reduction would be uncertain and would probably be small, particularly in the first half of the century.[27] However, significantly reducing U.S. greenhouse gas emissions without corresponding decreases in emissions generated by other large economies would probably not significantly reduce hurricane damage between now and 2075 because U.S. emissions make up a shrinking share of global emissions.

Of the two ways in which climate change is expected to increase hurricane damage—a rise in sea levels and an increase in the frequency of hurricanes—the first is the more certain; however, increases in sea levels between now and 2075 are expected to be relatively insensitive to changes in emissions over the same period. Experts find that once sea levels begin to rise, the process is hard to slow down.[28] Specifically, in the first half of the 21st century, the global increase in sea levels will be caused primarily by expansion of the oceans resulting from the warming of the water. That response is relatively insensitive to changes in emissions. Differences in emissions will begin to be more important in the second half of the century, when the melting of ice sheets is projected to play a more significant role.[29]

Table 4. Estimates of Federal Spending in 2075, Based on Likely Ranges of Expected Damage and Alternative Scenarios About Spending as a Percentage of Expected Damage

Likely Range of Federal Spending in 2075 if . . .	
Spending Equals 40% of Damage	
Measured as a percentage of GDP	0.06–0.13
Measured in billions of dollars based on today's economy	11–24
Spending Equals 60% of Damage	
Measured as a percentage of GDP	0.09–0.19
Measured in billions of dollars based on today's economy	16–34
Spending Equals 80% of Damage	
Measured as a percentage of GDP	0.12–0.25
Measured in billions of dollars based on today's economy	22–45

Source: Congressional Budget Office.

On the basis of historical data on federal spending as a percentage of economic damage for hurricanes occurring from August 2005 to the present, which averaged 60 percent, and on CBO's estimate of expected damage under current conditions, which is 0.16 percent of gross domestic product (GDP), expected federal spending under current conditions is 0.10 percent of GDP, or $18 billion.

Dollar amounts are expressed in 2015 dollars.

The "likely range" includes the middle two-thirds of the distribution of estimates from CBO's simulations.

Therefore, the IPCC projects that the likely range of the rise in sea levels by 2065 would be similar under a high-emissions scenario (wherein only limited efforts would be made to reduce emissions) or under a low-emissions scenario (wherein greenhouse gas emissions would be rapidly reduced starting immediately and completely eliminated by 2040). Reductions in global emissions would begin to have more significant effects on the rise in sea levels by the end of the century; but, even then, the extent of that rise under the low-emissions scenario is projected to be roughly two-thirds of that under the high-emissions scenario.[30]

Some of the factors affecting the frequency of hurricanes— including atmospheric temperatures, sea-surface temperatures, water vapor, and wind shear—respond relatively quickly to changes in emissions.[31] For example, the IPCC projects a significantly higher range of increases in the global mean surface temperature under the high-emissions scenario than under the low-emissions scenario.[32] Nonetheless, the relationship between climate change and hurricanes is not very well understood, which is why researchers project a wide range of changes in the frequency of hurricanes. Hence, the effect that a global reduction in greenhouse gas emissions would have on the frequency of hurricanes is unclear.

Moreover, even with U.S. efforts, stabilizing the concentration of greenhouse gases in the atmosphere would be virtually impossible if countries with rapidly growing economies, such as China, India, Brazil, and South Korea, did not substantially cut their emissions. Until about eight years ago, the United States generated more emissions than any other single country; China is now the single largest emitter. Whereas the United States currently accounts for about 15 percent of global emissions, China accounts for roughly 30 percent. The United States' share is projected to remain roughly constant over the next 25 years, while China's share is projected to grow to nearly 35 percent by 2040.[33]

Shift More Costs to State and Local Governments and Private Entities, Thereby Reducing Coastal Development

Providing federal assistance in the aftermath of hurricanes helps hard-hit communities, but it also reduces incentives for people to take into account the full impact on expected damage of their decisions about where to live and locate businesses. In essence, federal assistance subsidizes growth in areas that are vulnerable to hurricane damage by shielding households, businesses, and state and local governments in those areas from the financial consequences of such decisions. Requiring greater cost sharing by private entities and state and local governments would provide an incentive for them to more fully account for the extent of expected damage and thereby reduce the pace of coastal development.

Approaches for reducing the federal government's share of costs—and increasing the share borne by private entities and state and local governments—include:

- Increasing insurance requirements for households and businesses,
- Increasing the minimum amount of statewide per capita damage that is used as the primary criterion for providing federal assistance, and
- Reducing the statutorily set minimum share of costs borne by the federal government (as opposed to state and local governments) for assistance provided from FEMA's Disaster Relief Fund.

Increase Insurance Requirements

Administered by FEMA, the National Flood Insurance Program is the primary source of coverage for flood damage in the United States. Under the program, FEMA develops flood-hazard maps that indicate varying amounts of flood risk. Communities participating in the program become eligible for NFIP coverage, which is administered by private insurance companies, if they comply with building-code requirements and floodplain management practices established by FEMA. Moreover, residents who have a federally backed mortgage and live in an area where the annual flood risk is estimated to be at least 1 percent—called a special flood hazard area (SFHA)—are required to have flood insurance.

The existence of the NFIP, however, does not mean that businesses and households bear the full cost of flood risk. The federal government bears a significant share of the cost of such risk for two reasons. First, many property owners do not purchase flood insurance. A study conducted in 2006 by the RAND Corporation estimated that, nationwide, only 3 percent of single-family homes in communities participating in the NFIP had flood insurance, and only about half of the homes in SFHAs had insurance (in part because of incomplete compliance with the coverage requirement for properties with federally insured mortgages).[34] The availability of disaster assistance may affect people's decisions about purchasing flood insurance, but there is limited evidence about that relationship. One study found that for most households, the provision of federal assistance grants for damaged property did not affect the decision about whether to purchase flood insurance but did reduce the amount of insurance coverage that those households chose to purchase.[35]

Second, premium rates charged for insurance provided under the NFIP do not cover the full cost of insuring the property. About one-fifth of NFIP policies—generally those covering older structures—are explicitly subsidized, with the federal government covering about 50 percent to 55 percent of the premiums.[36] Moreover, subsidized policies are disproportionately subject to repetitive losses.[37] The Government Accountability Office (GAO) estimated that, in fiscal year 2013, the cost to the federal government of forgone premiums from subsidized policies was approximately $1.5 billion.[38] However, recent legislation (discussed below) will gradually phase out some of those subsidies through annual rate increases. In addition, some state insurance programs also subsidize policies covering hurricane risk.[39]

In addition, "full-risk" premium rates—that is, rates that are not explicitly subsidized—could underestimate the actual cost of insuring the properties. The flood maps that FEMA uses to set insurance rates may not accurately reflect changes in flood risk associated with development that has reduced wetlands and permeable ground area or with climate change that has begun to increase sea levels and may be changing the frequency and severity of hurricanes and rainstorms. Moreover, even when FEMA updates its maps, properties that might be classified as being in a higher-risk zone or at a lower elevation relative to expected flood heights are often "grandfathered" at their previous classifications.[40]

Explicit and implicit subsides to NFIP policyholders increase the likelihood that the premiums received will not cover claim payments, as occurred following Hurricanes Katrina and Sandy. Because the NFIP has had to borrow from the Treasury to pay claims, it now owes the Treasury about $23 billion.[41]

One option for decreasing the federal cost of hurricane damage is to set rates for coverage under the NFIP that would more closely parallel rates set by private insurers, while maintaining or strengthening requirements for homeowners and businesses to carry flood insurance. One study has found that greater coverage may reduce the need for disaster assistance, but the reduction is small.[42] However, the NFIP was designed to balance the goals of reducing the nation's long-term exposure to flood losses, making the program solvent, and meeting the statutory mandate to make the rates affordable. Balancing those objectives can be difficult. For example, raising premiums to more fully cover flooding risks would make it more difficult to expand participation in the flood insurance program. The NFIP was recently reauthorized through the Biggert-Waters Flood Insurance Reform Act of 2012 (subtitle A of Title II of Public Law 112-141), which included a number of reforms to address the future solvency and efficiency of the program. Those reforms included phasing out, and in some cases eliminating, subsidized premium rates for certain types of property in areas at high risk of flooding. However, the Homeowner Flood Insurance Affordability Act of 2014 (Public Law 113-89) reinstated certain premium subsidies and slowed down certain rate increases that had been mandated by the Biggert-Waters Act.[43]

Set a Higher Minimum Threshold for Disaster Declarations

Federal assistance provided through Stafford Act programs is authorized only when the President declares a major disaster in response to a request by a governor or tribal leader. Some analysts and policymakers maintain that such declarations have been made for incidents that could have been handled by the states.

Since 1986, FEMA has used the estimated amount of statewide damage relative to a state's population as an indicator of the burden of the disaster and generally relied on that indicator as a threshold in determining whether to recommend that the President declare a major disaster in a state. Although current law prohibits FEMA from making disaster determinations solely on the basis of population or income-based formulas, the statewide per capita indicator is the principal criterion FEMA uses for determining whether damage is sufficiently severe to warrant recommending a major disaster declaration.[44] That indicator, however, may underestimate a given state's capacity to recover from a disaster using its own resources. The Government Accountability Office reported in 2012 that the threshold for declaring a disaster is artificially low because it does not reflect the increase in per capita personal income that has occurred since the indicator was introduced in 1986 and because it only began to be adjusted upward for inflation in 1999.[45] As a result, the indicator understates the amount of economic resources available to states to fund their own responses to disasters. Adjusting the indicator upward, or setting some other more stringent requirement for measuring a state's need and capacity, could reduce the number of disaster declarations and, in turn, reduce federal spending on disaster assistance.

Another proposal that has been considered would require states to have their own disaster programs in place before they would be eligible to receive federal assistance under the Stafford Act. Such a requirement might increase states' capacity to handle emergencies without federal assistance.[46]

Reduce Federal Aid for Declared Disasters

The federal share of costs for relief activities funded by the Disaster Relief Fund (which has accounted for almost half of all federal spending on hurricane disaster relief since 2000) is designated by the Stafford Act to be at least 75 percent, with state and local governments funding the remaining 25 percent of disaster assistance; however, the federal share can increase if damage reaches certain thresholds.[47] Some analysts and policymakers have proposed that the federal share of costs could be reduced—to 50 percent, for example—whereas others argue that such an adjustment might place too great a burden on state and local governments.

Another approach that has been discussed would convert some or all of the federal assistance provided to states into low-interest or no-interest loans. Such loans could be structured to encourage states to plan for disasters—for example, by providing lower rates (or a larger share of assistance in the form of aid rather than loans) for states that have taken steps to prepare for disasters.

Setting limits on the circumstances for providing federal assistance—or on the share of costs borne by the federal government when a major disaster is declared—could reduce the amount of hurricane damage by providing state and local governments with greater incentives to put limits (such as zoning restrictions or stricter building-code requirements) on development in areas that are at relatively high risk of damage. Such limits could also increase the incentives of state and local governments to invest in infrastructure that is designed to reduce damage, such as seawalls.

Invest in Structural Changes to Reduce Vulnerability to Hurricane Damage

In recent years, federal agencies have placed greater emphasis on funding measures designed to reduce the vulnerability of property to future hurricane damage— typically referred to as hazard-mitigation measures.

Investments in hazard mitigation typically increase upfront costs of construction or restoration but reduce the costs associated with future damage.

One option for reducing federal expenditures on hurricane damage is to increase mitigation efforts. To the extent that the up-front costs of such measures are more than offset by the decrease in expected costs resulting from hurricane damage, the federal government could reduce total hurricane-related spending by undertaking more hazard mitigation or by providing incentives for state and local governments to do so.

In 2007, CBO examined a subset of mitigation efforts— FEMA's Pre-Disaster Mitigation Program—and concluded that projects undertaken as a result of that program reduced expected future losses (measured in discounted present value) by about $3 for each $1 spent on the projects.[48] Significant uncertainty surrounds that estimate, however, and the information available on past projects may not reliably indicate the effectiveness of additional mitigation projects in the future. Moreover, a complete assessment would need to consider whether mitigation measures, such as building seawalls, have the unintended effect of encouraging development in vulnerable areas.

FEMA lists a wide variety of hazard-mitigation measures that might be cost-effective (although assessing cost-effectiveness would require case-specific analyses). Such measures include elevating roads, redesigning or relocating bridges, enlarging culverts, elevating buildings, using flood-resistant construction materials, restoring and maintaining natural barriers such as sand dunes, installing hurricane clips between roof framing and walls, installing storm-resistant window shutters, and anchoring or repositioning rooftop equipment.[49]

APPENDIX A: ATTRIBUTING GROWTH IN EXPECTED HURRICANE DAMAGE TO CLIMATE CHANGE AND TO COASTAL DEVELOPMENT

The future increases in expected hurricane damage described in the main text of this report are the result of the combined forces of climate change (represented by changes in sea levels and in hurricane frequency) and coastal development (represented by changes in population and per capita income). The Congressional Budget Office attributes the increase in expected damage in 2075 to each of those forces.

CBO's results, as presented in the main text, express damage as a share of gross domestic product (GDP) because GDP is a measure of the nation's ability to pay for hurricane damage. To attribute growth in damage to the underlying forces of climate change and coastal development, however, it is necessary to measure damage on a dollar basis. Accordingly, in this section, CBO describes expected damage in constant 2015 dollars.

CBO's estimate of expected annual hurricane damage at present—the "reference case"—is based on estimates of the current frequency of hurricanes, current state-specific sea levels, and the current valuation of property exposed to hurricanes. According to CBO's estimates, mean expected hurricane damage in 2075 would be $150 billion, $120 billion more than the roughly $30 billion in expected damage that the agency estimates under current conditions. Because the $120 billion increase reflects the interaction of climate change and coastal development, that amount is larger than the sum of the increases that would occur if additional damage was the result only of climate change (holding property exposure—approximated by population and per capita income—at its current value) or if additional damage was the result only of coastal development (holding sea levels and hurricane frequencies at their current values). That is, the $120 billion increase includes the additional damage caused by climate change on the additional property exposure caused by coastal development. CBO attributes shares of that interaction to climate change and to coastal development (see Table A-1).

Climate Change Only

Accounting only for the effects of climate change, CBO estimates that mean expected damage in 2075 would be $65 billion, $35 billion higher than the estimate of expected damage under current conditions. The $65 billion was obtained by estimating expected damage using 2075 distributions for hurricane frequencies and sea levels, but holding population and per capita income constant at their levels in the reference case.

Coastal Development Only

Accounting only for the effects of coastal development, CBO estimates that mean expected damage in 2075 would be $70 billion, $40 billion higher than expected damage under current conditions. The $70 billion was obtained by estimating expected damage using the 2075 distributions for population and per capita income, but holding hurricane frequencies and sea levels constant at their levels in the reference case.

Climate Change and Coastal Development: Shares of Combined Effects

In combination, climate change and coastal development are estimated to increase hurricane damage by $45 billion more than the sum of the increase in damage that would occur as a result of each force on its own ($75 billion). That $45 billion reflects the *additional damage* that climate change has on the *additional property exposure* attributable to coastal development.

CBO allocates that $45 billion to climate change and to coastal development on the basis of the ratio of the increase in damage caused by each individual force on its own and the sum of the increase in damage caused by each of the two individual forces. For example, on its own, climate change is estimated to increase damage by $35 billion, or roughly 45 percent of the $75 billion sum of the increases in hurricane damage resulting from climate change only and from coastal development only. As a result, CBO allocated 45 percent of the $45 billion interaction effect to climate change ($20 billion) and the remaining 55 percent to coastal development ($25 billion).

On the basis of the allocation method described above, climate change accounts for $55 billion of the $120 billion increase in the average expected hurricane damage in 2075 (or 45 percent)—relative to CBO's reference case—and coastal development accounts for the remaining $65 billion (or 55 percent.)

Table A-1. Combined and Individual Effects of Climate Change and Coastal Development

	1	2	3	4	5
Source of Change in Damage	**Mean Damage Estimate in 2075 (Billions of dollars)**	**Increase in Mean Damage Estimate in 2075 Relative to $30 Billion in the Reference Case (Billions of dollars)[a]**	**Share of Sum of Individual Effects[b] (Percent)**	**Allocation of Interaction Effect (Billions of dollars)[c]**	**Contribution to Increase in Mean Damage Estimate of Total Combined Effect of Climate Change and Coastal Development (Billions of dollars)[d]**
Combined Effect of Climate Change and Coastal Development	150	120	n.a.	n.a.	n.a.
Individual Effect: Climate Change Only	65	35	45	20	55
Individual Effect: Coastal Development Only	70	40	55	25	65
Sum of Individual Effects	n.a. e	75	n.a.	n.a.	n.a.
Interaction Effect (Equals the combined effect minus the sum of individual effects)	n.a.	45	n.a.	n.a.	n.a.

Source: Congressional Budget Office, using data from Risk Management Solutions; Kerry A. Emanuel, "Downscaling CMIP5 Climate Models Shows Increased Tropical Cyclone Activity Over the 21st Century," *Proceedings of the National Academy of Sciences*, vol. 110, no. 30 (July 2013), www.pnas.org/content/110/30/12219; and Thomas R. Knutson and others, "Dynamical Downscaling Projections of Twenty-First-Century Atlantic Hurricane Activity: CMIP3 and CMIP5 Model-Based Scenarios," *Journal of Climate*, vol. 26, no. 17 (September 2013), http://journals.ametsoc.org/doi/abs/10.1175/JCLI-D-12-00539.1.

CBO's estimate of expected annual hurricane damage at present—the "reference case"—is based on estimates of the current frequency of hurricanes, current state-specific sea levels, and the current valuation of property exposed to hurricanes.

Dollar amounts are expressed in 2015 dollars and rounded to the nearest $5 billion. All shares are rounded to the nearest 5 percent. n.a. = not applicable.

[a] Equals the mean damage estimate reported in Column 1 minus the $30 billion of expected damage in CBO's reference case.

[b] Equals the increase in expected damage when individual effects are estimated in isolation ($35 billion for climate change only, for example) divided by the sum of the individual effects ($75 billion).

[c] Equals the share of the sum of individual effects caused by one factor reported in Column 3 (45 percent for climate change only, for example) multiplied by the interaction effect reported in Column 2 ($45 billion).

[d] Equals the sum of the individual increase in mean damage resulting from a single effect reported in Column 2 ($35 billion for climate change only, for example) plus the dollar amount of the interaction effect allocated to the individual effect reported in Column 4 ($20 billion for climate change only, for example).

[e] The mean individual effects cannot be summed because doing so would double count the $30 billion of damage in the reference case.

APPENDIX B: SOURCES USED BY CBO TO ESTIMATE FEDERAL SPENDING FOR HURRICANE DAMAGE

To estimate spending for each of the 16 major hurricanes that occurred between 2000 and 2015, the Congressional Budget Office relied on information from numerous sources.

- To estimate spending by the Federal Emergency Management Agency (FEMA) from the Disaster Relief Fund, which accounts for nearly half of all federal spending on damage from hurricanes, CBO drew from obligations data available in FEMA's monthly reports to the Congress, as well as obligations data available on FEMA's website.
- To estimate spending by the Department of Housing and Urban Development (HUD), CBO drew from obligations data provided by HUD that detail the disbursement of Community Development Block Grants to affected states, as well as information on HUD's website about those grants.

- To estimate spending by the Army Corps of Engineers for its civil works and disaster-response programs, CBO drew from obligations data and the Corps' spending plans for future projects.
- To estimate spending by the Small Business Administration for its disaster loans, CBO drew from loan disbursement data provided by that agency.
- To estimate spending by all other agencies, CBO drew from obligations data as available, the text of legislation and accompanying reports, and analyses of appropriations and agency spending produced by the Congressional Research Service.
- To estimate spending from FEMA's National Flood Insurance Program, CBO drew from claims payment data from FEMA's website.

In the case of recent major hurricanes for which there is an ongoing obligation of funds in some programs— specifically, for Sandy and Isaac (both of which occurred in 2012) and Irene (which occurred in 2011)—CBO used information on historic "spend-out" rates (that is, the rate at which agencies are expected to spend funds) and feedback from agencies to estimate ultimate amounts of spending. In cases in which federal assistance was provided to states affected by more than one hurricane or a different type of disaster, CBO used agency program documents and state implementation plans to assign spending to individual hurricanes in proportion to the amount of damage they caused.

End Notes

[1] See Terry Dinan, *CBO's Approach to Estimating Expected Hurricane Damage,* Working Paper 2016-02 (Congressional Budget Office, June 2016), www.cbo.gov/publication/51610.

[2] Those estimates are not related to CBO's 10-year baseline projection for disaster-related spending in the federal budget, which relies on different conventions. As directed by the law governing the construction of that baseline, federal funding for most discretionary disaster-related accounts in the federal budget is equal to the most recent year's annual appropriations for those accounts plus an increase to factor in the effects of inflation.

[3] See Terry Dinan, *CBO's Approach to Estimating Expected Hurricane Damage,* Working Paper 2016-02 (Congressional Budget Office, June 2016), www.cbo.gov/publication/51610.

[4] RMS is a catastrophe risk modeling company. Results produced by those models are used worldwide by many insurance companies, reinsurers, and financial institutions to assess their exposure to risk. RMS has developed damage functions that translate U.S. hurricane occurrences into state-specific estimates of expected damage. A more detailed discussion of

the damage functions and CBO's assessment of their validity is provided in the working paper that accompanies this report. For a description of the RMS model, see Michael Delgado and others, "Technical Appendix: Detailed Sectoral Models," in Trevor Houser and others, *American Climate Prospectus: Economic Risks in the United States* (Rhodium Group and Risk Management Solutions, October 2014), p. C-6, http://climate prospectus.org/publications.

[5] The state-specific projections that CBO used were supplied by RMS, which based its projections on Robert E. Kopp and others, "Probabilistic 21st and 22nd Century Sea-Level Projections at a Global Network of Tide-Gauge Sites," *Earth's Future*, vol. 2, no. 8 (August 2014; corrected, October 2014), pp. 383–406, http://onlinelibrary.wiley.com/doi/ 10.1002/2014EF000239/full. (The fact that the estimates were, by design, consistent with the IPCC's global projections was confirmed in a personal communication with Robert E. Kopp on September 17, 2015.)

[6] For a more complete discussion of the reasons that increases in sea levels will vary among locations, see ibid., pp. 383–384, http://onlinelibrary.wiley.com/doi/10.1002/2014 EF000239/full. See also National Oceanic and Atmospheric Administration, "Sea Level Trends: Frequently Asked Questions," *Tides and Currents* (revised October 15, 2013), http://tidesandcurrents.noaa.gov/ sltrends/faq.htm.

[7] For a discussion of these offsetting effects, see National Oceanic and Atmospheric Administration, Geophysical Fluid Dynamics Laboratory, *Large-Scale Climate Projections and Hurricanes* (updated December 8, 2015), www.gfdl.noaa.gov/global-warming-and-21stcentury-hurricanes.

[8] Thomas Knutson is a climate modeler at the U.S. Geophysical Fluid Dynamics Laboratory, a division of the National Oceanic and Atmospheric Administration. He is also an associate editor at the American Meteorological Society's *Journal of Climate* and is co-chair of the World Meteorological Organization Expert Team on Climate Change Impacts on Tropical Cyclones. Kerry Emanuel is a professor of meteorology at the Massachusetts Institute of Technology and specializes in the mechanisms acting to intensify hurricanes. He is a member of the U.S. National Academy of Sciences.

[9] The likely range in 2075 is from 0.15 percent of GDP to 0.31 percent. In 2050, it is from 0.14 percent to 0.26 percent.

[10] Hypothetically, if coastal development was the only condition affecting hurricane damage—that is, if sea levels and the frequency of hurricanes were held constant—damage would increase more slowly than GDP, CBO estimates. In particular, accounting only for changes in population and per capita income between now and 2075, CBO estimates, hurricane damage would increase at a rate that was roughly 60 percent less than the projected growth in GDP during that same period, resulting in expected damage in 2075 equal to roughly 0.10 percent of GDP.

[11] That attribution was based on the ratio between the increase in expected damage caused by the *individual factor* in isolation (climate change, holding population and per capita income constant and vice versa) to the *sum* of the increases in expected damage obtained when each factor was modeled in isolation. For additional information, see "Appendix A: Attributing Growth in Expected Hurricane Damage to Climate Change and to Coastal Development."

[12] That more-than-proportionate response has been estimated by some researchers. See Laura A. Bakkensen and Robert O. Mendelsohn, *Risk and Adaptation: Evidence From Global Hurricane Damages and Fatalities* (forthcoming).

[13] Agencies' roles are defined under the National Response Framework, which guides the nation's emergency response to a major disaster, and the National Disaster Recovery

Framework, which guides recovery activities. See Department of Homeland Security, *National Response Framework* (May 2013), http://tinyurl.com/gud8tw3; and Federal Emergency Management Agency, *National Disaster Recovery Framework: Strengthening Disaster Recovery for the Nation* (September 2011), http://go.usa.gov/cVMDd (PDF, 11 MB).

[14] See Bruce R. Lindsay, *FEMA's Disaster Relief Fund: Overview and Selected Issues*, Report for Congress R43537 (Congressional Research Service, May 7, 2014).

[15] See Jared T. Brown and Daniel J. Richardson, *FEMA's Public Assistance Grant Program: Background and Considerations for Congress,* Report for Congress R43990 (Congressional Research Service, April 16, 2015), p. 29.

[16] Because this report focuses on the budgetary implications of hurricanes, the estimates of expected damage do not take into account the loss of life caused by hurricanes.

[17] See Eugene Boyd, *Community Development Block Grant Funds in Disaster Relief and Recovery*, Report for Congress RL33330 (Congressional Research Service, November 1, 2012).

[18] See U.S. Army Corps of Engineers, "Emergency Operations: Disaster Response Missions and Information," www.usace.army.mil/Missions/EmergencyOperations.aspx.

[19] The $1.9 billion includes the cost of reducing interest rates on SBA loans and the expected costs associated with the probability that some amount of the loans would not be paid back.

[20] See Congressional Budget Office, *The Macroeconomic and Budgetary Effects of Hurricanes Katrina and Rita: An Update* (September 2005), www.cbo.gov/publication/17204.

[21] See Testimony of Douglas Holtz-Eakin, Director, Congressional Budget Office, *Macroeconomic and Budgetary Effects of Hurricanes Katrina and Rita* (October 6, 2005), www.cbo.gov/publication/ 17233. A recent working paper finds effects that extend over several decades, particularly for countries that experience frequent hurricanes. See Solomon M. Hsiang and Amir S. Jina, *The Causal Effect of Environmental Catastrophe on Long-Run Economic Growth: Evidence From 6,700 Cyclones*, Working Paper 20352 (National Bureau of Economic Research, July 2014), www.nber.org/papers/w20352.

[22] In the federal budget, those payments are classified as offsetting receipts, which are a credit against mandatory spending.

[23] See Congressional Budget Office, *The Macroeconomic and Budgetary Effects of Hurricanes Katrina and Rita: An Update* (September 2005), www.cbo.gov/publication/17204.

[24] In addition, following Katrina, lawmakers enacted the TANF [Temporary Assistance for Needy Families] Emergency Response and Recovery Act of 2005, which provided additional funds to states that were damaged by Hurricane Katrina or those that were hosting evacuees.

[25] The recent increase in the ratio of federal spending to total damage is a continuation of a longer-term trend. According to some researchers, federal assistance, measured as a proportion of hurricane damage, has grown significantly over the past 30 years and is dramatically larger than it was earlier in the 20th century. See J. David Cummins, Michael Suher, and George Zanjani, "Federal Financial Exposure to Natural Catastrophe Risk" in Deborah Lucas, ed., *Measuring and Managing Federal Financial Risk* (University of Chicago Press, 2010), pp. 61–92; and David A. Moss, "The Peculiar Politics of American Disaster Policy: How Television Has Changed Federal Relief," in Erwann Michel-Kerjan and Paul Slovic, eds., *The Irrational Economist: Making Decisions in a Dangerous World* (Public Affairs Books), pp. 151–160.

[26] Those estimates of hurricane-related spending are not related to CBO's 10-year baseline forecast for disaster-related spending in the budget. Whereas, for its 10-year baseline, CBO

projects federal spending from discretionary disaster-related accounts by inflating the most recent year's annual appropriations for those accounts, this report presents the annual demand for federal spending in a given year on the basis of the results of modeling expected damage. In its baseline for fiscal year 2016, CBO estimates federal spending from disaster-related accounts of approximately $10 billion. That figure is lower than the $18 billion in expected federal spending estimated in the analysis presented in this report (equal to roughly 60 percent of the $28 billion in expected damage under current conditions), in part because there has been little hurricane activity in the past few years and, therefore, smaller appropriations for disaster-related accounts. Conversely, if a storm comparable to Hurricane Sandy— which caused damage that resulted in lawmakers providing over $50 billion in supplemental appropriations—had struck in 2015, then CBO's baseline estimate of disaster-related spending would have exceeded the expected damage.

[27] In December 2015, nearly 200 nations pledged to keep global temperatures well below 3.6 degrees Fahrenheit above preindustrial times. Countries agreed to make nationally determined contributions (NDCs), which indicate their goals for reducing emissions, and to pursue domestic measures aimed at achieving those reductions. Every five years, countries are committed to reporting on progress in meeting their existing NDCs and to submit new ones.

[28] See Stefan Rahmstorf, "Sea Level in the 5th IPCC Report," *Real Climate* (October 15, 2013), www.realclimate.org/index.php/ archives/2013/10/sea-level-in-the-5th-ipcc-report.

[29] See Robert E. Kopp and others, "Probabilistic 21st and 22nd Century Sea-Level Projections at a Global Network of Tide-Gauge Sites," *Earth's Future*, vol. 2, no. 8 (August 2014; corrected, October 2014), pp. 383–406, http://onlinelibrary.wiley.com/doi/ 10.1002/2014EF 000239/full.

[30] For the period from 2046 through 2065, sea levels are projected to rise by between 0.6 and 1.1 feet under the scenario with the lowest emissions and by between 0.7 and 1.3 feet under the scenario with the highest emissions. For the period from 2081 through 2100, the rise in sea levels is projected to be between 0.9 and 1.8 feet (with a mean of 1.3 feet) under the scenario with the lowest emissions and between 1.5 and 2.7 feet (with a mean of 2.1 feet) under the scenario with the highest emissions. See Intergovernmental Panel on Climate Change, "Summary for Policymakers," in T.F. Stocker and others, eds., *Climate Change 2013: The Physical Science Basis. Contribution of Working Group 1 to the Fifth Assessment Report of the IPCC* (Cambridge University Press, 2013), Table SPM.2, p. 23, www.ipcc.ch/report/ar5/wg1.

[31] See Thomas Knutson and Robert E. Tuleya, "Impact of CO_2- Induced Warming on Simulated Hurricane Intensity and Precipitation: Sensitivity to the Choice of Climate Model and Convective Parameterization," *Journal of Climate*, vol. 17, no. 18 (September 2004).

[32] For the period from 2046 through 2065, the global mean surface temperature is projected to increase by 0.7 to 2.9 degrees Fahrenheit under the scenario with the lowest emissions and by 2.6 to 4.7 degrees Fahrenheit under the scenario with the highest emissions. See Intergovernmental Panel on Climate Change, "Summary for Policymakers," in T.F. Stocker and others, eds., *Climate Change 2013: The Physical Science Basis. Contribution of Working Group 1 to the Fifth Assessment Report of the IPCC* (Cambridge University Press, 2013), Table SPM.2, p. 23, www.ipcc.ch/report/ar5/wg1.

[33] See Energy Information Administration, *International Energy Outlook 2013* (July 2013), www.eia.gov/forecasts/archive/ieo13. For a general discussion of energy use and greenhouse gas emissions in China and their implications, see Andrew Stocking and Terry

Dinan, *China's Growing Energy Demand: Implications for the United States*, Working Paper 2015-05 (Congressional Budget Office, June 2015), www.cbo.gov/publication/50216.

[34] See Lloyd Dixon and others, *The National Flood Insurance Program's Market Penetration Rate: Estimates and Policy Implications* (RAND Corporation, February 2006), www.rand.org/pubs/ technical_reports/TR300.html.

[35] The study found that, in the case of disaster loans provided by the Small Business Administration—which, unlike disaster grants, must be repaid by recipients—the size of a loan had no effect on the amount of insurance purchased. The authors also note that increased reliance on HUD's CDBG Disaster Recovery Program may also reduce insurance coverage because some states have used the flexibility provided under the program to give homeowners larger grants to repair property than would be allowed under FEMA's Individual Assistance Program. Further, the authors caution that because most federal disaster assistance is provided to local governments, such assistance may reduce local governments' incentives for investing in hazard mitigation. For more information, see Carolyn Kousky, Erwann O. Michel-Kerjan, and Paul A. Raschky, *Does Federal Disaster Assistance Crowd Out Private Insurance?* Working Paper (Wharton Risk Center, 2014), http://tinyurl.com/gu83bls (PDF, 235 KB).

[36] Subsidized policies cover certain properties in areas at high risk of damage from flooding—primarily those built before flood insurance rate maps (FIRMs) became available and, thus, without a clear understanding of their vulnerability to flooding. In 2013, FEMA estimated that people and businesses holding such policies were paying premiums that were 45 percent to 50 percent of their full-risk value. For more information, see Congressional Budget Office, *The National Flood Insurance Program: Factors Affecting Actuarial Soundness* (November 2009), p. 6, www.cbo.gov/ publication/41313.

[37] Ibid., p. 24.

[38] See Government Accountability Office, *Flood Insurance: Forgone Premiums Cannot Be Measured and FEMA Should Validate and Monitor Data System Changes*, GAO-15-111 (December 2014), p. 25, www.gao.gov/products/GAO-15-111.

[39] See Kent Smetters and David Torregrosa, *Financing Losses From Catastrophic Risks*, Working Paper 2008-09 (Congressional Budget Office, November 2008), pp. 32–33, www.cbo.gov/ publication/20400.

[40] In some cases, the grandfathering does not adversely affect total premiums paid to the NFIP but instead raises the rates paid by other policyholders, resulting in cross-subsidies within the program. See Congressional Budget Office, *The National Flood Insurance Program: Factors Affecting Actuarial Soundness* (November 2009), pp. 14–17, www.cbo.gov/ publication/41313.

[41] See TreasuryDirect, Department of the Treasury, "Federal Borrowings Program Reports: Detail Principal and Accrued Interest Balances and Summary General Ledger Balances (All Funds)," Fund 70x4236 (January 2016), www.treasurydirect.gov/ govt/reports/tbp/ tbp_2016_01.htm.

[42] Ibid.

[43] As a result of the Biggert-Waters Act and the Homeowner Flood Insurance Affordability Act, FEMA is phasing out subsidies for the following property-rate classes by implementing annual rate increases of 25 percent: any property that has experienced severe repetitive losses; nonprimary residences; business properties; any property that has incurred flood-related damage for which the cumulative amount of NFIP payments has equaled or exceeded the property's fair-market value; and any property that has sustained substantial damage or had substantial improvement exceeding 50 percent of the property's fair-market

value. FEMA is phasing out subsidies for the following property-rate classes by implementing annual rate increases of 5 percent to 15 percent: primary residences purchased after July 6, 2012, and primary residences not insured by the NFIP as of July 6, 2012. FEMA has eliminated subsidies for NFIP policies that have lapsed and for any policies for which the insured refuses to accept any offer of mitigation assistance following a major disaster or in connection with a repetitive loss. For a more detailed discussion of the cost of forgone premiums from subsidized policies, see Government Accountability Office, *Flood Insurance: Forgone Premiums Cannot Be Measured and FEMA Should Validate and Monitor Data System Changes*, GAO-15-111 (December 2014), www.gao.gov/products/ GAO-15-111.

[44] Section 320 of the Stafford Act prohibits FEMA from denying Stafford Act assistance to an area solely on the basis of a mathematical formula or sliding scale that is based on income or population. See Bruce R. Lindsay, *FEMA's Disaster Relief Fund: Overview and Selected Issues,* Report for Congress R43537 (Congressional Research Service, May 7, 2014), p. 16.

[45] Originally set at $1.00 per person in 1986, FEMA's statewide per capita indicator was $1.35 in 2012, having been periodically adjusted for inflation since 1999. GAO estimated in 2012 that the indicator would have increased to $3.57 in 2011 had it been adjusted for increases in per capita income since 1986 and would have increased to $2.07 in 2012 had it been adjusted for inflation since 1986. GAO estimated that, of disaster declarations made from 2004 to 2011, 44 percent would not have met the threshold to qualify for assistance if the statewide indicator had been adjusted for per capita income, and 25 percent would not have qualified if the indicator had been adjusted for inflation since 1986. For more information, see Government Accountability Office, *Federal Disaster Assistance: Improved Criteria Needed to Assess a Jurisdiction's Capability to Respond and Recover on Its Own*, GAO-12-838 (September 2012), www.gao.gov/products/GAO12-838.

[46] For a more detailed discussion of these options, see Bruce R. Lindsay and Justin Murray, *Supplemental Appropriations for Disaster Assistance: Summary Data and Analysis*, Report for Congress R43665 (Congressional Research Service, October 1, 2014).

[47] See Bruce R. Lindsay, *FEMA's Disaster Relief Fund: Overview and Selected Issues*, Report for Congress R43537 (Congressional Research Service, May 7, 2014), p. 20.

[48] Present value is a single number that expresses a flow of future income (or payments) in terms of an equivalent lump sum received (or paid) today.

[49] See Federal Emergency Management Agency, *Hurricane Mitigation: A Handbook for Public Facilities* (May 2005), www.fema.gov/media-library/assets/documents/16562 .

In: Hurricane Damage
Editor: Martha Atkinson
ISBN: 978-1-53610-035-8

Chapter 2

CBO'S APPROACH TO ESTIMATING EXPECTED HURRICANE DAMAGE*

Terry Dinan

ABSTRACT

This working paper describes how the Congressional Budget Office estimates the effects of climate change and coastal development on hurricane damage. The estimates themselves are presented in a separate report—*Potential Increases in Hurricane Damage in the United States: Implications for the Federal Budget*—for three selected future years: 2025, 2050, and 2075.

Climate change is projected to increase damage in two ways. First, climate change is projected to result in more frequent high-intensity hurricanes. Second, for any given storm, rising sea levels are projected to lead to increased damage from storm surges. CBO generates state-specific estimates of hurricane damage on the basis of existing property exposure (which corresponds to existing vulnerability-weighted populations and per capita income in each state) by using damage functions provided by Risk Management Solutions and estimates of the distributions of hurricane frequencies and state-specific sea levels in future years.

Coastal development is also projected to increase damage simply by putting more people and property in harm's way. In this analysis, coastal

* This is an edited, reformatted and augmented version of Working Paper 2016-02, issued by the Congressional Budget Office, June 2016.

development is measured as changes in population and per capita income in areas that are vulnerable to hurricane damage. Specifically, CBO inflates those state-specific damage estimates on the basis of each state's distributions of vulnerability-weighted population and per capita income in future years, as well as on elasticities that translate changes in population and per capita income into changes in the magnitude of damage.

In its June 2016 report *Potential Increases in Hurricane Damage in the United States: Implications for the Federal Budget*, the Congressional Budget Office estimated hurricane damage in future years. Details on the approach—known as a Monte Carlo method—that CBO used to develop underlying distributions for hurricane frequencies, sea levels, population, and per capita income, along with a more expanded description of how CBO estimated expected damage on the basis of those inputs, are described in this paper.

Overview of CBO's Estimation Process

CBO estimated a distribution of hurricane damage by simulating damage 5,000 times, with each simulation, n (n = 1 to 5,000), based on a unique set of values for changes in the frequency of hurricanes and for state-specific estimates of sea level, population, and per capita income selected from distributions for a future year.

Twenty-two states—all of which CBO estimated to have a nonzero probability of incurring hurricane damage—were included in the agency's model. Because growth in some regions (along the coast, for example) will have a larger effect on damage than growth in other regions, measures of population and per capita income were weighted on the basis of their relative vulnerability to hurricane damage, with $\ddot{p}$ and $\ddot{y}$ indicating vulnerability-weighted population and per capita income, respectively, and p and y indicating unweighted values.

The values for hurricane frequencies, f, sea levels, s, vulnerability-weighted population, $p\ddot{p}$, and vulnerability-weighted per capita income, $\ddot{y}$, in turn, were each selected from individual distributions in each specific future year: 2025, 2050, or 2075. The shape of CBO's damage distribution in a particular year, such as 2075, depends on the shape of the 2075 distributions for f, s, $\ddot{p}$, and $\ddot{y}$ and on the relationship between those variables and hurricane damage.

The distributions of hurricane frequencies and sea levels that CBO used were estimated by university or government researchers (or by CBO, using data provided by those researchers). The distributions of vulnerability-weighted population and per capita income were developed by CBO. For each of the three future years (t = 2025, 2050, and 2075), CBO selected hurricane frequencies from 18 sets of expected frequencies, where each set included a value for each hurricane Category c, $c = 1$ (for a Category 1 hurricane) through $c = 5$ (for a Category 5 hurricane) for each year; the selection probabilities for both hurricane frequencies and sea levels in the simulations are described below. (There are five categories of hurricanes, which are classified on the basis of their peak wind speed, with Category 5 storms being the most intense.) The other variables discussed here ($\ddot{p}$ and $\ddot{y}$) have normal distributions.

CBO compared distributions of expected damage in each future year with an estimate of expected damage in a reference case. For the reference case, hurricane frequencies, f, were based on estimates for 2010, and all other variables, s, $\ddot{p}$, and $\ddot{y}$, were set at their estimated values for 2015.[1] For notational convenience throughout this paper, the t subscript is suppressed when denoting future years. Subscripts i, j, and k are used to indicate county, state, and region, respectively; subscript n indicates that the variable takes on a different value in each nth simulation; and subscript R indicates that the variable is set at its reference value. Thus, for example, $S_{j,}$ denotes sea level in state j in the nth simulation, and S_{jR}denotes sea level in state j in the reference case. For general purposes, a damage estimate for state j can be described as $D_j\left(f_x, s_{j,x}, \ddot{p}_{j,x}, \ddot{y}_{j,x}\right)$, where $x = R$ indicates that Dj was calculated with the variable set at its reference value, and $x = n$ indicates that Dj was calculated with the variable set at its value selected in the nth simulation.

Each simulation of CBO's model begins with a set of draws for all four of the conditions that affect expected hurricane damage (see Figure 1). Each nth simulation of the model determines a set of state-specific estimates of expected damage (reflecting only the effects of climate change) based on the draws for hurricane frequency, f, and sea levels, s, in that simulation; existing property exposure in each state; and a set of damage functions developed by Risk Management Solutions (RMS). Those climate-only damage estimates are then adjusted to reflect the effects of coastal development. That adjustment is based on draws of each county's population and per capita income in 2075—which are weighted to reflect the county's relative vulnerability to damage from wind and storm surges and then aggregated to the state level (creating variables $\ddot{p}$ and $\ddot{y}$)—along with state-specific inflation factors developed by CBO.

For each simulation, *n*, values of the four random variables *f*, *s*, $\ddot{p}$, and $\ddot{y}$ were drawn from their individual distributions, and those variables were used to estimate expected damage for each state *j* (*j* = 1 through 22). The *n*th damage estimate (corresponding to the *n*th simulation) for state *j* is:

$$D_j(f_n, s_{j,n}, \ddot{p}_{j,n}, \ddot{y}_{j,n}) = \sum_{c=1}^{5} f_n(c) d_{j,n}(c, s_{j,n}, \ddot{p}_{j,R}, \ddot{y}_{j,R}) g_{j,n}(\ddot{p}_{j,n}, \ddot{y}_{j,n})$$

where

- $d_{j,n}(c, s_{j,n}, \ddot{p}_{j,R}, \ddot{y}_{j,R})$ is the expected damage in dollars in state *j*, given U.S. landfall of a hurricane of Category *c*, the specific value of sea level for state *j* selected for the *n*th simulation, and state *j*'s population and per capita income in the reference case (reflecting state *j*'s property exposure in 2015); and
- $g_{j,}(\ddot{p}_{j,n}, \ddot{y}_{j,n})$ is a damage inflation factor. It increases $d_{j,}(c, S_{j,n}, \ddot{p}_{j,R}, \ddot{y}_{j,R})$ on the basis of the estimates of state *j*'s vulnerability-weighted population and per capita income in year *t* as selected in the *n*th simulation. As described below, each state's population and per capita income can be affected by rising sea levels.

The damage inflation factor, $g_{j,n}$, depends on the change in population and per capita income in each state (relative to 2015) and a set of state-specific population and per capita income elasticities (indicating the percentage change in expected damage given a percentage change in population or per capita income) developed by CBO (see Figure 2). Specifically,

$$g_{j,n}(\ddot{p}_{j,n}, \ddot{y}_{j,n}) = 1 + \Delta\ddot{p}_{j,n}\epsilon_j^p + \Delta\ddot{y}_{j,n}\epsilon_j^y$$

where:

$\ddot{p}_{j,n}$ = the vulnerability-weighted population of state j in the nth simulation

$\Delta\ddot{p}_{j,n}$ = $\frac{\ddot{p}_{j,n}}{\ddot{p}_{j,R}} - 1$

$\ddot{p}_{j,R}$ = the vulnerability-weighted population of state j in the reference case

ϵ_j^p = the percentage change in expected damage in state j given a percentage change in population in state j

$\ddot{y}_{j,n}$ = the vulnerability-weighted per capita income value for state j in the nth simulation

$\Delta\ddot{y}_{j,n}$ = $\frac{\ddot{y}_{j,n}}{\ddot{y}_{j,R}} - 1$

$\ddot{y}_{j,R}$ = the vulnerability-weighted per capita income of state j in the reference case

ϵ_j^y = the percentage change in expected damage in state j given a percentage change in per capita income in state j.

Total expected damage in the United States corresponding to the nth simulation is obtained by aggregating across the 22 state damage estimates for that simulation:

$$D_n = \sum_{j=1}^{22} D_j\left(f_n, s_{j,n}, \ddot{p}_{j,n}, \ddot{y}_{i,n}\right)$$

For each selected year (2025, *2050,* and *2075),* this process is repeated 5,000 times to generate a distribution of expected hurricane damage in the United States.

CBO compared distributions of expected future damage with a reference case, which is the estimate of expected damage obtained by setting all variables at their reference levels (denoted by subscript *R):*

$$D_R = \sum_{j=1}^{22} D_j\left(f_R, s_{j,R}, \ddot{p}_{j,R}, \ddot{y}_{i,R}\right)$$

Damage Functions

CBO projects the magnitude of expected hurricane damage by using damage functions provided by RMS.[2] Those functions estimate expected damage on a state-specific basis, given:

- Existing exposure of residential and nonresidential property in the state,
- Landfall of a specific category of hurricane (Categories 1 through 5*)* anywhere in the United States, and
- State-specific estimates of sea levels.

Those estimated losses account for the probability that the state will incur no losses when a hurricane of a particular category makes landfall in the United States. For example, if a Category *5* hurricane was to make landfall in the United States, it would be much more likely to strike the southern section of the United States' eastern coast than the northern section. As a result, the estimated expected damage would be much smaller in New Jersey (roughly *$15* million under current conditions) than in Florida (roughly $1.8 billion under current conditions). If a Category *5* hurricane actually made landfall in New Jersey, the losses would be very large; however, the relatively small *expected* loss reflects the small probability of that occurring.

Estimating thc probability that a hurricane of a particular category will make landfall at any given location is difficult given the infrequency with which hurricanes occur, particularly the most damaging Category 4 and 5 storms. RMS addressed that problem by using more than 100,000 simulations of hurricane seasons under current conditions (with frequencies of simulated hurricanes constrained to the frequencies observed over the past 100 years and with hurricanes following physically realistic pathways).[3]

CBO assessed the validity of using damage functions provided by RMS in this analysis by comparing RMS's damage estimates for actual hurricanes that have occurred since 2002 with estimates generated by the National Oceanic and Atmospheric Administration (NOAA).[4] For this purpose, RMS modeled the specific storms by using estimates of property exposure at the time the hurricane occurred. RMS estimated exposure in previous years by adjusting downward the monetary value of current property exposure in its model to account for trends in development between the time of landfall and the present.[5]

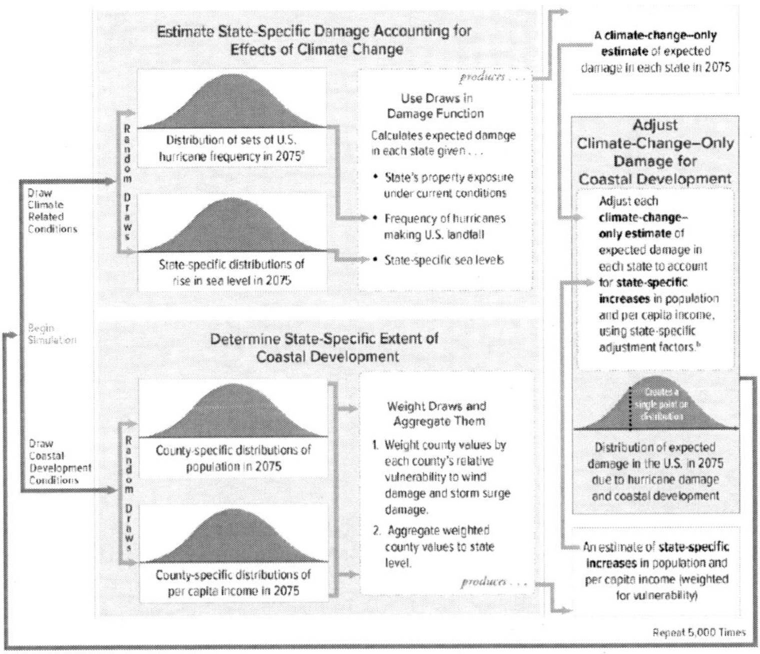

Source: Congressional Budget Office.

[a] Each set consists of a projection of frequency for hurricanes in each of five categories. (The five categories of hurricanes are based on peak wind speed. Category 5 storms are the most intense.)

[b] Each state's increase in expected damage due to an increase in its population and per capita income is uniquely determined based on the share of the state's expected damage (measured under current conditions) that comes from wind versus storm-surge damage. That unique determination incorporates different responses of wind and storm-surge damage to a given increase in population and per capita income.

Figure 1. Flow of the Model for Estimating the Effects of Climate Change and Coastal Development on Hurricane Damage in Selected Future Years: Example Year, 2075.

For individual storms, some of RMS's estimates were higher than NOAA's (most significantly for Hurricane Katrina); however, on average, RMS's estimates were lower—equal to 80 percent of NOAA's estimates. Excluding Hurricane Katrina from the calculation, RMS's estimates were, on average, 2 percent higher than NOAA's. In the case of Katrina, RMS's method for adjusting property exposure is not able to replicate the significant changes in exposure in New Orleans as a direct result of Hurricane Katrina itself, and consequently the downward adjustment underestimates property exposure in New Orleans in 2005.[6]

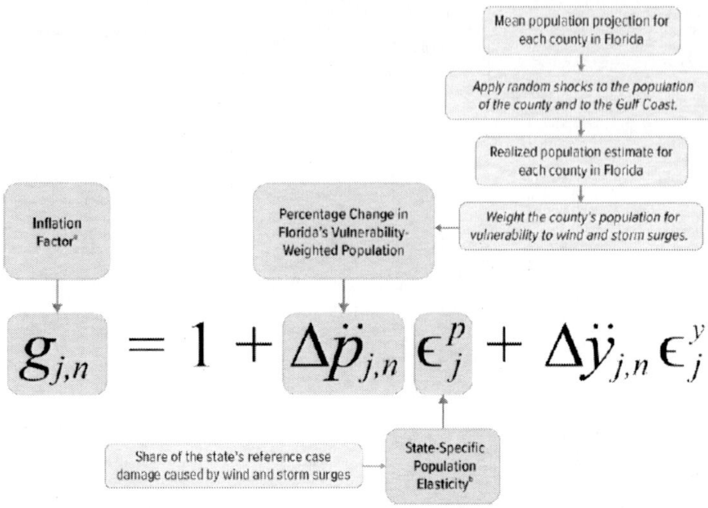

Source: Congressional Budget Office.

CBO constructed a measure of the percentage change in each state's vulnerability-weighted per capita income, Δyÿjj,n, by using the same method presented in this figure for the percentage change in vulnerability-weighted population. The agency also constructed a state-specific per capita income elasticity, Ejj yy, which indicates the percentage change in damage given a percentage change in per capita income.

[a] Inflation factor is used to adjust the estimate of expected climate-only damage for the effects of coastal development.

[b] Population elasticity indicates the percentage change in damage given a percentage change in population.

Figure 2. Estimating Effects of Climate Change and Coastal Development in 2075: Example, Florida.

FREQUENCY OF HURRICANES

The estimated effects of climate change on the frequency of various categories of hurricanes depend on changes in the climatic conditions affecting hurricane formation (changes in sea surface temperatures, for example) as well as the relationship between those conditions and the occurrence of hurricanes. CBO uses 18 different sets of predictions about the frequency of hurricanes—with each set providing a prediction of the annual frequency of each of the five categories of hurricanes. Those 18 sets include predictions that are based on the following: different concentrations of greenhouse gases in the atmosphere that correspond to different emission scenarios and land-use patterns, different models that link such concentrations to changes in the conditions that cause hurricanes, and different models that predict hurricanes on the basis of changes in those conditions.[7]

The 18 sets of frequency projections that CBO uses include 11 sets that were constructed using a downscaling model developed by Thomas Knutson and 7 sets constructed using a downscaling model constructed by Kerry Emanuel.[8] (The downscaling models estimate regional effects on the basis of output from global climate models.) To avoid having the 11 sets of hurricane frequencies produced by Knutson be more influential than the 7 sets produced by Emanuel—simply because there are more of them—CBO drew from each researcher's hurricane frequencies with a probability of 0.5 for its simulations. Specifically, the probabilities are about 4.5 percent (0.5/11) for each of Knutson's sets and about 7 percent (0.5/7) for each of Emanuel's sets.

Knutson and Emanuel both predict hurricane frequencies in future years on the basis of projections of factors that influence storms (such as sea surface temperature and wind shear in the Atlantic Basin). Those projections are derived from a number of different coupled atmosphere-ocean general circulation models (AOGCMs). Those AOGCMs were used in the Coupled Model Intercomparison Project (CMIP), an undertaking in which all models were run using the same set of analyses (for example, making projections on the basis of a specific concentration of greenhouse gases in the atmosphere and reporting results over the same time frames). As a result, CMIP isolates differences in climate outcomes resulting from differences in the models that project such outcomes, rather than differences in the scenarios that the researchers modeled.

Emanuel projected landfalls of hurricanes in the United States on the basis of the results of six individual AOGCMs that were used in the fifth (most recent) CMIP (CMIP5) as well as the "CMIP5 ensemble," which projects landfalls on the basis of hurricane-influencing factors that are obtained by averaging the results of each AOGCM.[9] Knutson estimated hurricane occurrences in the North Atlantic by using projections from the CMIP5 ensemble as well as results from 10 individual AOGCMs used in an earlier phase of the CMIP.[10] CBO translated Knutson's basin-level hurricane projections into U.S. landfalls on the basis of a matrix provided by RMS. For example, that matrix indicates the probability that a Category 4 hurricane that forms in the North Atlantic Basin will make landfall in the United States as a Category 4, 3, 2, or 1 hurricane or that it will diminish to a tropical storm.

Emanuel and Knutson's hurricane projections were derived using different assumptions about concentrations of greenhouse gases in the atmosphere. Specifically, Emanuel used AOGCM results that were based on an assumption of higher concentrations of greenhouse gases in the atmosphere than the model results that Knutson used. Emanuel's landfall projections were based on Representative Concentration Pathway (RCP) 8.5—a concentration of greenhouse gases in the atmosphere projected under a scenario in which both emissions and the conversion of terrain to cropland or pastureland continues to increase over the next century.[11] The global surface temperature, averaged between 2081 and 2100, is projected to increase under the RCP8.5 scenario by 6.7°F (in relation to the average preindustrial temperature).[12] Knutson's projections were based on RCP4.5, which is a concentration that could occur if emissions were to peak in 2040 and then begin to decline after that, and if less terrain was converted to cropland and pastureland than under the RCP8.5 scenario. Researchers estimate that, averaged between 2081 and 2100, global surface temperature would increase by 3.24°F under the RCP4.5 scenario.[13]

RISING SEA LEVELS

As the global climate warms, sea levels rise because of the thermal expansion of seawater and the melting of ice sheets in Greenland and Antarctica. The most recent report by the Intergovernmental Panel on Climate Change (IPCC) concluded that sea levels are rising globally, that the rate at which they are rising has increased since preindustrial times, and that the rate will accelerate in this century.[14] Rising sea levels increase damage caused by storm surges; thus, estimated damage from any given category of hurricane increases as sea levels rise. CBO's analysis takes that effect into account by using state-specific predictions of changes in sea level for the three selected future years (2025, 2050, and 2075).

The predictions CBO used were based on data provided by RMS. Specifically, RMS provided estimates at nine specified percentiles of the distributions of rising sea levels for each state and for each decade; CBO interpolated to obtain values for sea level values for 2025 and 2075.[15] (The estimates differ for different states for several reasons, including nonuniform changes in ocean dynamics, heat content, and salinity, as well as variation in the rates of vertical land motion attributable to factors such as tectonics and the withdrawal of local groundwater and hydrocarbons.)[16] The probabilities

CBO attached to each of the nine percentiles are shown in Table 1; for example, the 66.7th percentile was chosen with a probability of 0.172, or 17.2 percent of the time. For each simulation, the same percentile was used for all the states.

In turn, RMS based its percentile distributions of rising sea levels by state and decade on predictions provided by climate scientist Robert Kopp.[17] Those predictions were based on alternative assumptions about the concentration of greenhouse gases in the atmosphere (known as representative concentration pathways, or RCPs)—and about changes in rising sea levels for any RCP.[18] For example, the percentiles for rising sea levels combine potential outcomes associated with each of three different RCPs used by the IPCC: RCPs 2.6, 4.5, and 8.5.[19] (As described above, each scenario corresponds to a unique set of assumptions about emissions and land-use patterns.)

Table 1. Percentiles and Corresponding Probabilities of Rising Sea Levels

Percentile Observation for Rising Sea Levels	Probability of Drawing the Percentile Observation
0.5	0.017
5.0	0.078
16.7	0.146
33.3	0.172
50.0	0.171
66.7	0.172
83.3	0.146
95.0	0.078
99.5	0.017

Source: Congressional Budget Office.

On a global scale, the predictions from Kopp and his colleagues are similar to those found in other assessments. For example, 90 percent of the projections by Kopp and others for the global rise in sea levels by 2100 are between 1 ft. and 4 ft. Similarly, the IPCC's *Fifth Assessment Report* finds that 90 percent of the projections for 2100 lie between 1.2 ft. and 3.2 ft.[20]

VULNERABILITY-WEIGHTED POPULATION ESTIMATES FOR EACH STATE

In its analysis, CBO used projections of population and per capita income as a proxy for property exposure, a method that is consistent with previous research on hurricane damage.[21] CBO first estimated population at the county level. Because growth in some counties (those along the coast, for example) will have a larger effect on the state's expected damage than growth in others (inland ones, for example), CBO weighted each county on the basis of its relative vulnerability to hurricane damage. Those vulnerability-weighted county estimates were then aggregated to the state level. CBO also allows for increases in sea level that substantially increase hurricane damage to slow growth in population (and per capita income, as discussed below).

Estimates of County Population

CBO's model incorporated 777 counties, including all counties that were found to have a nonzero probability of incurring hurricane damage (described below). For each simulation, CBO used a county population estimate that was based on a mean projection and on both a regional shock and a county shock, such that the shocks affecting counties within a region had a joint normal distribution (see Figure 3).

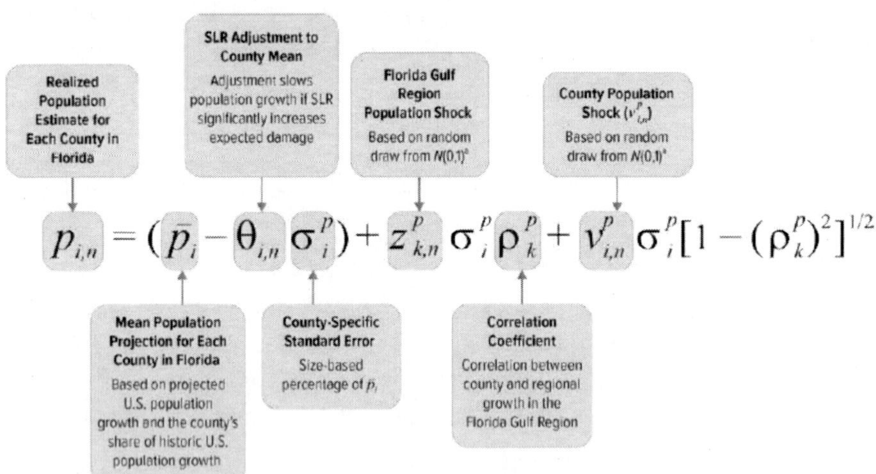

Source: Congressional Budget Office. SLR = sea level rise.

[a] $N(0,1)$ is a standard normal distribution, with a mean of zero and a standard deviation of 1.

Figure 3. Applying Random Shocks to Generate County Population Estimates for Each Simulation: Example, Florida.

CBO adjusted county means on the basis of potential increases in damage in a given state resulting from rising sea levels (allowing significant increases in sea levels) and the associated hurricane damage in that state (to slow the county's population growth):

$$p_{i,n}\left(\theta_{i,n}, z^{p}_{k,n}, v^{p}_{i,n}\right) = (\bar{p}_i - \theta_{i,n}\sigma^{p}_{i}) + z^{p}_{k,n}\sigma^{p}_{i}\rho^{p}_{k} + v^{p}_{i,n}\,\sigma^{p}_{i}[1 - (\rho^{p}_{k})^2]^{1/2}$$

where

$p_{i,n}$	=	county i's estimated population projection in the nth simulation
$\theta_{i,n}$	=	an adjustment to county i's mean population projection on the basis of the extent to which rising sea levels in the nth simulation are estimated to increase damage in the state in which county i resides (see a more detailed description of $\theta_{i,n}$ below)
$z^{p}_{k,n}$	=	the draw for the population shock for region k (in which county i resides) in the nth simulation (the shock is obtained from a standard normal distribution)
$v^{p}_{i,n}$	=	the draw for the population shock for county i in the nth simulation, obtained from a standard normal distribution
$\bar{p}_i$	=	county i's mean population projection (a fixed value estimated by CBO, as described below)
σ^{p}_{i}	=	the standard deviation of county i's population distribution (a fixed value that is equal to $x\bar{p}_i$, where $x = 0.1$, 0.11, or 0.12, as described below)
ρ^{p}_{k}	=	the correlation between the historical population growth rate of region k and the population-weighted growth rates for the individual counties within the region (a fixed value estimated by CBO, as described below).

Projections of Mean County Population Growth (Estimates of $\bar{p}_i$). For all but 26 of the 777 counties, CBO projected their population growth between 2010 and 2040 on the basis of their historic population growth between 2000 and 2010 relative to that of the total U.S. population over the same period. For example, if a county accounted for 1 percent of the growth in the total U.S. population between 2000 and 2010, then CBO estimated that the county would account for 1 percent of the growth in the U.S. population over the forecast period. The total U.S. population growth is based on CBO's macroeconomic forecast.[22] That method preserves the underlying variation in counties' growth rates while ensuring that the county-specific projections are consistent with CBO's aggregate population projection.[23]

CBO chose to project counties' populations on the basis of their growth rates between 2000 and 2010 rather than over a longer historical period because the 777 counties grew very rapidly (relative to the U.S. population as a whole) in the second half of the 20th century, and CBO judged that such a trend was unlikely to continue.[24] Measured between 1950 and 2000, the population-weighted growth rate of the 777 counties in this analysis was more than three times higher than the growth rate of the U.S. population as a whole. Measured over the decade from 2000 to 2010, the populations of the 777 counties still grew faster than that of the United States as a whole, but only 22 percent faster. The relatively rapid population growth of the 777 counties between 1950 and 2000 was fueled, in part, by a significant southern migration prompted by the increased availability of air conditioning; consequently, that growth is unlikely to continue.

For the other 26 counties, CBO instead used county-specific population projections that were made for regional planning purposes.[25] Those projections were made by county or city planning departments, state governments, or state universities. Those 26 counties include:

- The 20 counties with the largest populations in 2010. Specific projections were obtained for those counties since the most populated counties tend to make up a relatively large share of their state's estimated damage.
- Six counties that had populations of more than 100,000 and had a difference of two or more percentage points between the average annual growth over the 1950–2010 period and the 2000– 2010 period. Those two criteria captured counties, such as the parishes surrounding New Orleans, that had unusual circumstances between 2000 and 2010. (New Orleans experienced a sharp decline in population after Hurricane Katrina in 2005.)

Given the difficulty of knowing whether each coastal county will continue to grow faster or slower than the U.S. total population over the long run, CBO estimates that for 2040 and beyond all 777 counties will grow at the same rate that CBO projects for the United States as a whole.

Uncertainty About Projections of County Population (Estimates of $\sigma^p_{i,n}$*).* Given the uncertainty about population growth, CBO estimated each county's future population as a normal distribution with a mean estimated by the process described above and a standard deviation equal to a percentage of its population. Specifically, CBO estimated standard deviations equal to:

- 10 percent of the population for counties with populations greater than 100,000
- 11 percent of the population for counties with populations of 50,000 to 100,000
- 12 percent of the population for counties with populations less than 50,000

Those estimates of standard deviations are based on a study conducted by Stanley K. Smith and others.[26] Standard deviations are likely to be larger for smaller cities because a given change in population (for example, if the opening of a new manufacturing plant attracted 5,000 new residents) corresponds to a larger share of the existing population of a small city than of a larger one.

Correlation in Growth Between Counties and the Region in Which They Reside (Estimates of ρ^{p}_{k}*).* CBO's estimates of each county's population include its own population shock and a regional shock, each of which are determined by random draws from a standard normal distribution. That method was chosen after exploring the extent to which adjoining counties' decade-specific growth rates—that is, growth during each decade from 1950 to 2010—were correlated, and whether growth rates in adjoining states were correlated. (CBO's examination of correlation between adjoining counties focused primarily on Florida, which accounted for more than half of estimated damage in the reference case.) That analysis revealed no clear pattern of correlation in growth between adjoining counties within a state but indicated distinct patterns of correlation between growth for clusters of states along the Gulf and East Coasts.

On the basis of that analysis, CBO defined four regions for the purpose of projecting population growth and estimated a correlation coefficient, $p^{P}{}_{k}$, for each region:[27]

- Florida Gulf (Alabama, Florida, Louisiana, Mississippi, and Texas)— $p^{P} = 0.287$
- Southern Coastal (Georgia, North Carolina, and South Carolina)—$p^{P} = 0.184$
- Mid-Atlantic and Northern (Connecticut, Delaware, Maryland, Massachusetts, New Jersey, New York, Pennsylvania, Rhode Island,

Virginia, and West Virginia, as well as Washington, D.C.)—p^P = 0.149

- Far Northern (Maine, New Hampshire, and Vermont)—pP = 0.469.

Adjustment to County Population Means on the Basis of Rising Sea Levels (Estimates of $\theta_{j,n}$*).* CBO accounts for the potential for rising sea levels—and the resulting rise in expected hurricane damage from storm surges—to slow population growth in vulnerable states. Specifically, CBO adjusted each county's estimated mean population on the basis of the estimated increase in expected damage from storm surges for the state in which that county resides. The quantitative effect of expected damage associated with rising sea levels (or even of actual hurricane damage) on population growth is unknown. The adjustment used here incorporates a threshold effect: A rise in sea level must increase the state's expected hurricane damage by at least 25 percent before its counties' population means are adjusted. The adjustment also has an upper bound: It cannot reduce mean population estimates by more than a specified amount, set here at 1 standard deviation from the unadjusted mean.

The adjustment factor, è*jjᴀnn*, reduces the county's mean population estimate, *pp̄ii*, if the rise in sea level in the state *j* (in which county *i* resides) increases state *j*'s damage in the *n*th simulation by more than 25 percent relative to its damage in the reference case. For each state *j*:

$$\theta_{j,n}$$

$$= 0;\ \text{if } \Delta\hat{d}_{j,n} \leq 0.25,$$

$$= \min(1, \Delta\hat{d}_{j,n});\ \text{if } \Delta\hat{d}_{j,n} > 0.25,$$

where

$$\Delta\hat{d}_{j,n} = \frac{D_j\left(f_R, s_{j,n}, p_{j,R}, y_{j,R}\right)}{D_j\left(f_R, s_{j,R}, p_{j,R}, y_{j,R}\right)} - 1$$

For example:

$$\text{if } \Delta\hat{d}_{j,n} = 0.5, \text{ then } \theta_{j,n} = 0.5$$

$$\text{if } \Delta\hat{d}_{j,n} = 1.2, \text{ then } \theta_{j,n} = 1.$$

On the basis of that adjustment factor, county *i*'s mean population, $p\bar{p}ii$, would be set at 1 standard deviation below the unadjusted mean if the sea level draw in the *n*th simulation (holding all other variables at their reference levels) led to at least a doubling of estimated damage in the state in which county *i* is located. For example, if the draw for sea levels in the *n*th simulation increased expected damage in Florida by at least 25 percent (relative to Florida's expected damage in the reference case), then the mean population estimates of all the counties in Florida would be reduced in that *n*th simulation.

ESTIMATES OF VULNERABILITY-WEIGHTED COUNTY POPULATIONS

Development in each state will probably increase the damage caused by a given storm; however, the effect on hurricane damage depends on where the development occurs. Development in counties that are relatively vulnerable to hurricane damage will increase their state's damage estimate more than development in counties that are relatively invulnerable to such damage. To measure the effect of each county's development on its state's estimated damage, CBO weighted each county's growth in population and per capita income on the basis of its vulnerability to damage from storm surges and wind damage:

$$\ddot{p}_{i,n} = p_{i,n}[\lambda_i(1 - w_j) + \gamma_i w_j]$$

where

$\ddot{p}_{i,n}$	=	vulnerability-weighted population of county *i* in the *n*th simulation
λ_i	=	the weight used to indicate vulnerability of county *i* (in state *j*) to storm surge damage relative to all other counties in state *j*
$(1 - w_j)$	=	share of damage in state *j* that comes from storm surges (as opposed to wind)
γ_i	=	the weight used to indicate vulnerability of county *i* (in state *j*) to wind damage relative to all other counties in state *j*
w_j	=	share of state *j*'s damage that comes from wind (as opposed to storm surges).

Surge Damage Weights. CBO's surge damage weight for each county *i*, in state *j*, is equal to the probability-weighted loss ratio from storm surges in county *i*, relative to the total of such probability-weighted losses, summed across all counties in state *j* (in which county *i* resides):

$$\lambda_i = \frac{\sum_{c=1}^{5} m_i(c) q_j(c)}{\sum_{i=1}^{I_j} \sum_{c=1}^{5} [m_i(c) q_j(c)]}$$

where

$m_i(c)$	=	storm-surge loss ratio in county *i* (in state *j*), given that a hurricane of Category *c* imposes losses on state *j*
$q_j(c)$	=	probability that a hurricane of Category *c* occurs and imposes losses on state *j*. CBO used estimates of $q_j(c)$ generated by RMS's Hurricane Model (see above description)
I_j	=	the number of counties in state *j*.

The maximum potential total building losses in county *i*—given an occurrence of a Category *c* hurricane—divided by estimates of the total value of the buildings in the county is equal to $mm_i(cc)$. CBO calculated each county's loss ratio by using data from the Federal Emergency Management Agency's (FEMA) Coastal Flood Loss Atlas (CFLA), version 3.0, which FEMA developed using the Hazus loss estimation model.[28]

In essence, the weight $\ddot{e}ii$ is county *i*'s share of the total increase in probability-weighted damage from storm surges that state *j* would experience if an additional $1 of property was added to each county in the state. For example, the weight for Rockingham, New Hampshire, is 0.79, indicating that it accounts for 79 percent of the total additional expected storm-surge damage that New Hampshire would incur if $1 of additional property was added to each county in the state. In contrast, Hillsborough, New Hampshire (which is land-locked), has a zero weight, indicating that adding more property to Hillsborough would not increase expected storm-surge damage in New Hampshire. Surge weights for all the counties in any given state sum to one; that is, $\sum_{i=1}^{I_j} \lambda_i = 1$.

State *j*'s share of damage that comes from storm surges, as opposed to wind—$(1 - w_j)$—is calculated on the basis of data provided by RMS that indicates the breakdowns of state-specific damage in the reference case. Each state's total damage is attributed either to storm-surge damage or to wind damage.

Wind Damage Weights. CBO's wind damage weight for each county *i,* in state *j,* is equal to the probability-weighted wind loss ratio in county *i,* relative to the total of such probability-weighted losses, summed across all counties in state *j*:

$$\gamma_i = \frac{\sum_{c=1}^{5} h_i(c) q_j(c)}{\sum_{i=1}^{I_j} \sum_{c=1}^{5} [h_i(c) q_j(c)]}$$

where

$h_i(c)$ = loss ratio due to wind damage in county *i* of state *j*, given U.S. landfall of a hurricane of Category *c*.

CBO used two sources in generating estimates of *h*(*c*). Maps of sustained surface wind speeds produced by the National Hurricane Center (NHC) were used to identify the maximum winds each county would be expected to experience if a hurricane in Category *c* made landfall along its state's coastline.[29] (To calculate that maximum, CBO assumed that the hurricane made landfall at the section of the coast closest to the county.) Relationships between wind speed and damage were derived from FEMA's Hazus loss-estimation model. Those relationships, termed wind loss ratios, indicate a county's maximum building damage as a share of its total building valuations for a given wind speed.[30] The estimated ratios indicate that losses increase more than proportionately as wind speeds increase; for example, Category 3 wind speeds, averaging 120 mph, resulted in an average loss ratio of 0.13;

Category 4 wind speeds, averaging 143 mph, resulted in an average loss ratio of 0.44.

As was the case for the surge weight, the wind weight calculated for county *i*, ãii, is equal to *i*'s share of the total increase in state *j*'s probability-weighted wind damage that would occur if $1 of additional property was added to each county in the state. For example, Rockingham, New Hampshire, had a wind weight of 0.35, indicating that it would account for 35 percent of the total increase in expected wind damage that New Hampshire would experience under those circumstances. Wind weights for all the counties in any given state sum to 1; that is, $\sum_{i=1}^{I_j} \gamma_i = 1$

AGGREGATING VULNERABILITY-WEIGHTED COUNTY POPULATION ESTIMATES TO THE STATE LEVEL

Each state's vulnerability-weighted population is simply the sum of the vulnerability-weighted populations of the counties within it:

$$\ddot{p}_{j,n} = \sum_{i=1}^{I_j} \ddot{p}_{i,n}$$

where

$\ddot{p}_{j,n}$ = the vulnerability-weighted population of state *j* in the *n*th simulation.

ESTIMATES OF VULNERABILITY-WEIGHTED STATE PER CAPITA INCOME

CBO projected each state's per capita income by using the same method that was used to project the state's population. Counties' per capita incomes were projected on the basis of a mean projection and both regional- and county-level shocks, such that the shocks affecting counties within a region have a joint normal distribution. Mean per capita income estimates were adjusted on the basis of increases in hurricane damage, and county estimates were weighted on the basis of their relative vulnerability to wind and storm-surge damage as well as the state's share of damage from wind and storm surges. County estimates were aggregated to obtain state totals.

ESTIMATES OF COUNTY PER CAPITA INCOME

For each simulation, CBO calculated a unique estimate of per capita income for each county:

$$y_{i,n}\left(\theta_{i,n}, z^{y}_{k,n}, v^{y}_{i,n}\right) = (\bar{y}_i - \theta_{i,n}\sigma^{y}_{i}) + z^{y}_{k,n}\sigma^{y}_{i}\rho^{y}_{k} + v^{y}_{i,n}\sigma^{y}_{i}[1 - (\rho^{y}_{k})^2]^{1/2}$$

where

$y_{i,n}$	=	county *i*'s estimated per capita income projection in the *n*th simulation
$\theta_{i,n}$	=	an adjustment to the projection of county *i*'s mean per capita income on the basis of the extent to which sea level rise in the *n*th simulation is estimated to increase damage (same factor used for $p_{i,n}$)
$z^{y}_{k,n}$	=	the draw for the per capita income shock to region *k* (containing county *i*) for the *n*th simulation of the model, obtained from a standard normal distribution
$v^{y}_{i,n}$	=	the draw for the per capita income shock for county *i* in the *n*th simulation, obtained from a standard normal distribution
$\bar{y}_i$	=	county *i*'s mean population projection
σ^{y}_{i}	=	the standard deviation of county *i*'s per capita income distribution (a fixed value equal to $0.11\bar{y}_i$, as described below)
ρ^{y}_{k}	=	the correlation between the historical per capita income growth rate of region *k* and the population-weighted population growth rates for the individual counties within the region *R*.

Mean Projections (Estimates of $\bar{y}_{i,n}$*).* Because county-level projections of per capita income growth through 2075 are not available, CBO modeled each county's mean per capita income as growing (through 2040) at a weighted average of:

- Its growth rate between 1990 and 2000 (the decade preceding the recession),
- Its growth rate between 2000 and 2010, which reflects the effects of the recession and is the most recent decade for which census data are available, and

- The growth rate CBO projected for the United States as a whole.

Specifically, each of the county's two historic growth rates was assigned a weight of 0.1, and the U.S. growth rate was assigned a weight of 0.8. That method allows each county's historic growth to influence its future growth but also ensures a degree of consistency between the growth rates of the counties included in this analysis and the rate of growth that CBO projects for the United States as a whole. As was the case with estimates of population growth, CBO projected that, after 2040, each county's per capita income would grow at the same rate as that projected for the United States as a whole.

Uncertainty About Projections of per Capita Income (Estimates of **of $\sigma^y_{i,n}$).** CBO estimated a distribution of county-level per capita income by using a standard deviation set at 11 percent of each county's mean per capita income; that is, $\sigma_{i,n} = 0.11\overline{y_{i,n}}$. Because the economic literature does not provide estimates of the uncertainty surrounding county-level income projections, CBO chose to use the same standard deviation that was used for estimates of the population growth of midsized counties (those with populations of 50,000 to 100,000).

Correlation in Growth Between Counties and the Region in Which They Reside (Estimates of ρ^y_k**).** CBO estimated correlation coefficients between the growth in counties' per capita income and growth in the per capita income of the region in which they reside. That correlation was estimated for the same four regions used in estimating population growth (described above). Specifically, CBO estimated ρ^y_k for each region:[31]

- Florida Gulf—ρ^y = 0.727
- Southern Coastal—ρ^y = 0.794
- Mid-Atlantic and Northern—ρ^y = 0.504
- Far-Northern—ρ^y = 0.810.

Adjustment to Counties' Mean per Capita Income on the Basis of Rising Sea Levels. CBO adjusted each county's mean per capita income by using the same adjustment variable, $\theta_{j,n}$, that was used to adjust its mean population.

As described above, that adjustment incorporates a threshold effect: A rise in sea level must increase a state's hurricane damage by at least 25 percent before its counties' mean per capita incomes are affected. The adjustment also has an upper limit: If the draw for rising sea levels in the *n*th simulation causes a state's damage to double (or more than double), its counties' mean per capita incomes used in the *n*th simulation will be decreased by only 1 standard deviation.

ESTIMATE OF VULNERABILITY-WEIGHTED PER CAPITA INCOME FOR EACH COUNTY

As was the case with estimates of county population, CBO weighted each county's per capita income to reflect the county's relative vulnerability to damage from storm surges and from wind, as well as the fraction of the state's total damage that is attributed to storm surges versus wind. Specifically, CBO estimated each county's vulnerability-weighted per capita income as:

$$\ddot{y}_{i,n} = y_{i,n}[\lambda_i(1 - w_j) + \gamma_i w_j]$$

where

$\ddot{y}_{i,n}$	=	vulnerability-weighted per capita income of county *i* in the *n*th simulation
λ_i, γ_i, w_j	=	the same weights that are used to estimate vulnerability-weighted population of county *i* in the *n*th simulation.

Aggregating Estimates of Counties' Vulnerability-Weighted per Capita Income to the State Level

Projections of vulnerability-weighted per capita income for each state *j* are calculated as the sum of the estimates of population- and vulnerability-weighted county per capita income:

$$\ddot{y}_{j,n} = \sum_{i=1}^{I_j} \frac{\ddot{p}_{i,n}}{\ddot{p}_{j,n}} \ddot{y}_{i,n}$$

where:

$\ddot{y}_{j,n}$ = the vulnerability-weighted per capita income in state j in the nth simulation.

Elasticities

Estimates of hurricane damage are sensitive to assumptions about how much hurricane damage will increase in response to increases in population and per capita income in vulnerable areas. Information on the elasticity of hurricane damage with respect to socioeconomic variables—that is, the percentage change in damage given a percentage change in population or per capita income—is limited.[32] Some studies, such as one conducted by Roger A. Pielke and others, assume an elasticity of 1, implying that a doubling of both income and population would lead to a quadrupling of damage. Similarly, in a study published in 2010, William Nordhaus assumed that damage is proportionate to gross domestic product.[33] In contrast, Robert Mendelsohn and others estimated the global income elasticity of damage as 0.42 and the population damage elasticity as –0.2.[34] More recent work by Laura Bakkensen and Robert Mendelsohn found that the United States had significantly different elasticities from those of other countries:[35]

- *Responses to Increases in Population.* Bakkensen and Mendelsohn found no evidence that increases in population led to increases in wind damage in the United States (an elasticity of 0). In contrast, they found evidence of adaptation by countries in the Organization for Economic Cooperation and Development (OECD; excluding the United States) with respect to population growth (a statistically significant elasticity of less than 1).
- *Responses to Increases in Per Capita Income.* Bakkensen and Mendelsohn estimated that hurricane damage in the United States increased more than proportionately to increases in per capita income (an elasticity estimate of 1.15). In contrast, they found a statistically significant per capita income elasticity of less than 1 for non-OECD countries.

The elasticities estimated in the studies described above reflect both intentional and unintentional changes in the vulnerability of communities to damage as population and per capita income increase. For example, as population increases, buildings may be constructed closer to each other (reducing vulnerability to wind damage) or the use of multistory housing may increase (reducing the vulnerability to storm surges). Likewise, as per capita income increases, buildings may be better built, owners may take measures to reduce vulnerability, or communities may invest in coastal protection by constructing levees.[36]

CBO allowed wind and storm-surge damage to each have unique responses to changes in population and per capita income. CBO's elasticity estimates were informed by the estimates of Bakkensen and Mendelsohn (the only study that estimated elasticities explicitly for the United States). However, CBO modified Bakkensen and Mendelsohn's elasticities because the reliability of their estimates is likely to be limited by the size of their data set: They had 110 observations for the United States.[37] As described in this working paper's companion report, CBO examined the sensitivity of its damage estimates to the elasticities used in this analysis.

- *Storm-Surge Damage*. CBO estimated that storm-surge damage would increase less than proportionately to increases in per capita income (an elasticity of 0.75) and to increases in population (an elasticity of 0.5).
- *Wind Damage*. CBO estimated that wind damage would increase proportionately with increases in per capita income (an elasticity of 1) and less than proportionately with increases in population (an elasticity of 0.25).

CBO anticipates that damage from storm surges would increase less in response to increases in per capita income (an elasticity of 0.75) than would damage from wind (an elasticity of 1.0) because higher per capita income could motivate either public or private entities to invest in infrastructure (such as seawalls) that is designed to limit damage from storm surges. In contrast, CBO expects that damage from storm surges would respond more to increases in population (an elasticity of 0.5) than would wind damage (an elasticity of 0.25). Unlike with wind damage, increases in population density would not necessarily provide protection from storm-surge damage: Increasing the number of single-story buildings in a given area would probably result in a proportional increase in the amount of damage from storm surges. However, increases in damage from storm surges would be less than proportional to increases in population to the extent that increased population led to the construction of taller buildings or of public infrastructure designed to limit damage from storm surges. To reflect the fact that potential differences in storm damage would depend on whether population growth led to the construction of more single-story homes or taller buildings, CBO used an elasticity of 0.5.

According to the elasticities described above, doubling both population and per capita income would result in a 125 percent increase in damage from both wind and storm surges. A specific set of elasticities is calculated for each state on the basis of the shares of today's estimated damage that are caused by wind and storm surges. Thus,

$$\epsilon_j^p = 0.5(1 - w_j) + 0.25w_j$$

$$\epsilon_j^y = 0.75(1 - w_j) + w_j$$

where

ϵ_j^p	=	percentage increase in damage in state j, given a percentage increase population
ϵ_j^y	=	percentage increase in damage in state j, given a percentage increase in per capita income
$(1 - w_j)$	=	the share of total damage in state j that comes from storm surges
w_j	=	the share of total damage in state j that comes from wind.

End Notes

[1] This reference case of estimated expected damage under current conditions is a more appropriate comparison to expected future damage than actual damage in any particular year for the following reasons: Actual hurricane damage may be unusually high or low depending on whether the number of hurricanes in each category making landfall in that year was higher or lower than average and whether landfalls occurred in densely or sparsely populated areas. Likewise, the distribution of actual hurricane damage in any selected future year would be wider than the distribution of expected damage that CBO estimates.

[2] For a description of this model, see Michael Delgado and others, "Technical Appendix: Detailed Sectoral Models," in Trevor Houser and others, American Climate Prospectus: Economic Risk9 in the United States (Rhodium Group and Risk Management Solutions, October 2014), p. C-6, http://climateprospectus.org/publications/. Damage estimates include direct damage to property and contents caused by wind and storm surges, as well as indirect damage caused by interrupted business activity.

[3] See Michael Delgado and others, “Technical Appendix: Detailed Sectoral Models,” in Trevor Houser and others, American Climate Prospectus: Economic Risks in the United States

(Rhodium Group and Risk Management Solutions, October 2014), p. C-6, http://climateprospectus.org/publications/.

[4] NOAA's method of estimating damage is described in Adam B. Smith and Richard W. Katz, "U.S. Billion-Dollar Weather and Climate Disasters: Data Sources, Trends, Accuracy, and Biases," Natural Hazards, vol. 67, no. 2 (June 2013), Table 3, pp. 387–410, http://dx.doi.org/10.1007/s11069-013-0566-5.

[5] This information was provided to CBO by RMS for the purpose of making this comparison. These comparisons were made using RMS's estimate of "ground-up wind and full surge losses," which is the measure that CBO used in its analysis.

[6] Paul Wilson, Risk Management Solutions, personal communication (March 29, 2015).

[7] Land use affects the stock of carbon stored in vegetation. For example, turning forestland into cropland releases carbon that had been stored in the trees and the soil.

[8] See Kerry A. Emanuel, "Downscaling CMIP5 Climate Models Shows Increased Tropical Cyclone Activity Over the 21st Century," Proceedings of the National Academy of Sciences, vol. 110, no. 30 (July 2013), pp. 12219–12224, www.pnas.org/content/110/30/12219; and Thomas R. Knutson and others, "Dynamical Downscaling Projections of TwentyFirst-Century Atlantic Hurricane Activity: CMIP3 and CMIP5 Model-Based Scenarios," Journal of Climate, vol. 26, no. 17 (September 2013), pp. 6591–6617, http://journals.ametsoc.org/doi/abs/10.1175/JCLI-D-12-00539.1.

[9] The hurricane projections that Emanuel based on the CMIP5 ensemble results are shown in Kerry A. Emanuel, "Downscaling CMIP5 Climate Models Shows Increased Tropical Cyclone Activity Over the 21st Century," Proceedings of the National Academy of Sciences, vol. 110, no. 30 (July 2013), pp. 12219–12224, www.pnas.org/content/110/30/12219. The results from the downscaling of individual AOGCM models were obtained directly from the author and have not yet been published.

[10] Knutson's method and the CMIP5 ensemble results are shown in Thomas R. Knutson and others, "Dynamical Downscaling Projections of Twenty-First-Century Atlantic Hurricane Activity: CMIP3 and CMIP5 Model-Based Scenarios," Journal of Climate, vol. 26, no. 17 (September 2013), http://journals.ametsoc.org/doi/abs/10.1175/JCLI-D-12-00539.1. Although Knutson did not project hurricane occurrences on the basis of the output of the individual AOGCM models used in CMIP5, he did so for 10 individual models used in CMIP3 (the third phase of the CMIP). On the basis of advice provided by Knutson, CBO used the percentage variations found between the downscaling of individual CMIP3 model results and downscaling CMIP3 ensemble results to build an equivalent amount of variation around the CMIP5 ensemble.

[11] See Detlef P. van Vuuren and others, "The Representative Concentration Pathways: An Overview," Climatic Change, vol. 109, no. 1 (November 2011), pp. 5–31, http://link.springer.com/article/10.1007%2Fs10584-011-0148-z.

[12] See Intergovernmental Panel on Climate Change, "Summary for Policymakers," in T.F. Stocker and others, eds., Climate Change 2013: The Physical Science Basis. Contribution of Working Group I to the Fifth Assessment Report of the IPCC (Cambridge University Press, 2013), Table SPM.2, p. 23, www.ipcc.ch/report/ar5/wg1/.

[13] Ibid.

[14] See Intergovernmental Panel on Climate Change, "Sea Level Change," Chapter 13 in T.F. Stocker and others, eds., Climate Change 2013: The Physical Science Basis. Contribution of Working Group I to the Fifth Assessment Report of the IPCC (Cambridge University Press, 2013), Figure 13.27, p. 1204, www.ipcc.ch/report/ar5/wg1/.

[15] RMS provided 11 percentiles but had damage functions corresponding to only nine of the percentiles.

[16] See Robert E. Kopp and others, "Probabilistic 21st and 22nd Century Sea-Level Projections at a Global Network of Tide-Gauge Sites," Earth's Future, vol. 2, no. 8 (August 2014; corrected, October 2014), pp. 383–406, http://onlinelibrary.wiley.com/doi/10.1002/2014EF000239/full; and National Oceanic and Atmospheric Administration, Tides and Currents, "Frequently Asked Questions" (October 15, 2013), http://tidesandcurrents.noaa.gov/sltrends/faq.htm.

[17] In particular, Kopp provided decade-specific percentile estimates for 79 locations defined by latitude and longitude. RMS then identified the states corresponding to those values. (Paul Wilson, Risk Management Solutions, personal communication, August 30, 2015). The analysis by Kopp and others combines potential outcomes associated with three scenarios about the concentration of greenhouse gases, called representative concentration pathways (RCPs), used by the IPCC: RCPs 2.6, 4.5, and 8.5. Global sea level rise through 2050 is caused primarily by thermal expansion of the ocean and does not differ greatly in the three scenarios. Differences in the RCPs are more important in the second half of the century, when the melting of global ice sheets plays a more significant role.

[18] Global sea level rise through 2050 is caused primarily by thermal expansion of the ocean and is relatively insensitive to changes in emissions. Differences in RCPs begin to be more important in the second half of the century, when the melting of global ice sheets plays a more significant role. See Robert E. Kopp and others, "Probabilistic 21st and 22nd Century Sea-Level Projections at a Global Network of Tide-Gauge Sites," Earth's Future, vol. 2, no. 8 (August 2014; corrected, October 2014), pp. 383–406, http://onlinelibrary.wiley.com/doi/10.1002/2014EF000239/full.

[19] For RCPs 2.6, 4.5, and 8.5, the IPCC predicts an increase in global surface temperature, averaged between 2081 and 2100 (and measured relative to pre-industrial levels), of 1°C, 1.8°C, and 3.7°C, respectively. See Intergovernmental Panel on Climate Change, "Summary for Policymakers," in T.F. Stocker and others, eds., Climate Change 2013: The Physical Science Basis. Contribution of Working Group 1 to the Fifth Assessment Report of the IPCC (Cambridge University Press, 2013), Table SPM.2, p. 23, www.ipcc.ch/report/ar5/wg1/.

[20] The IPCC's "likely" range for sea level rise encompasses 90 percent of the distribution and is compared with the "very likely" range estimated by Kopp and others, which also encompasses 90 percent of the distribution. Although the IPCC projects sea level rise for ranges of years, CBO has used the IPCC's projections of global mean sea level rise in 2100 to best compare with Kopp's results. See Intergovernmental Panel on Climate Change, "Sea Level Change," chap. 13 in T.F. Stocker and others, eds., Climate Change 2013: The Physical Science Basis. Contribution of Working Group I to the Fifth Assessment Report of the IPCC (Cambridge University Press, 2013), Table 13.5, p. 1182, www.ipcc.ch/report/ar5/wg1/.

[21] For further discussion, see Laura A. Bakkensen and Robert O. Mendelsohn, "Risk and Adaptation: Evidence From Global Tropical Cyclone Damages and Fatalities," Journal of the Association of Environmental and Resource Economics (forthcoming).

[22] See Congressional Budget Office, The 2015 Long-Term Budget Outlook (June 2015), www.cbo.gov/publicationl50250.

[23] This method is called the "share of growth, apportionment method"; see Stanley K. Smith, Jeff Tayman, and David A. Swanson, State and Local Population Projections: Methodology and

Analysis (Kluwer Publications, 2002), p. 179, www.springer.com/us/book/978O3O64 64928.

[24] Althou counties' population growth rates between 2000 and 2010 were influenced by the downturn in the economy during that period, the selection of that decade is unlikely to bias projections made through 2040. In particular, the use of the 2000-2010 growth rates would bias CBO's projections of county growth only if the downturn systematically reduced growth in the 777 counties more, or less, than it reduced population growth in the United States as a whole.

[25] CBO identified sources for county-specific projections from the Census Bureau's list of state-level offices that manage population estimates. See Census Bureau, Federal—State Cooperative for Population Estimates, "FSCPE Contacts," www.census.gov/popest/fscpe/coop.html.

[26] See Stanley K. Smith, Jeff Tayman, and David A. Swanson, State and Local Population Projections: Methodology and Analysis (Kluwer Academic Publishers, 2002), Table 13.3, p. 317, www.springer.com/us/book/9780306464928.

[27] The values for pc, were obtained by regressing each county's decade-specific population growth rate (for each decade between 1950 and 2010) against the decade-specific population growth rate for the region in which the county resides. Each county's decade-specific growth rate was weighted by its population in that decade.

[28] Version 3.0 of FEMA's CFLA combines the National Hurricane Center's SLOSH model, which models storm-surge heights, with FEMA's Hazus model, which is a regional multihazard loss-estimation model. CBO used an output attribute (C#_BLDG_LR) from the CFLA for building loss ratios. Those loss ratios represent total damage to buildings divided by actual building valuations for each county modeled in a worst-case "maximum of maximums" storm-surge scenario. Although those loss ratios are based on worst-case scenarios, they are useful for identifying each county's relative contribution to the potential damage that could occur in its state. For more information on the CFLA and the Hazus-MH Coastal Flood Model, see H.E. Longenecker and others, "Hazus-MH Coastal Flood Model: FEMA Region IV Standard Operating Procedure for Coastal Flood Hazard and Loss Analysis" (FEMA Region IV, updated August 2012), http://tinyurl.com/zde3hos (PDF, 13.4 MB).

[29] NHC's maps of sustained surface wind speeds are available online at National Oceanic and Atmospheric Administration, National Hurricane Center, "The Inland Wind Model and the Maximum Envelope of Winds," (January 20, 2016), www.nhc.noaa.gov/aboutmeow.shtml.

[30] CBO used building loss ratios generated by running the Hazus Hurricane Model and selecting only for wind damage. The loss ratios sustained in a particular location were correlated with the maximum wind speeds experienced at that location to produce a wind-speed-to-damage curve. For more information on the Hazus Hurricane Model and wind damage curves, see Department of Homeland Security, Federal Emergency Management Agency, "Hurricane Model Technical Manual," Hazus—MH 2.1: Technical Manual (January 2015), www.fema.gov/media-library/assets/documents/24609.

[31] The values for $\rho k\ y$ were obtained by regressing each county's decade-specific per capita income growth rate (for each decade between 1960 and 2000) against the decade-specific per capita income growth rate for the region in which the county resides. Each county's decade-specific growth rate was weighted by its population in that decade.

[32] See Laurens Bouwer, "Projections of Future Extreme Weather Losses Under Changes in Climate and Exposure," Risk Analysis, vol. 33, no. 5 (May 2013), pp. 915–930, www.ncbi.nlm.nih.gov/pubmed/22958147.

[33] See Roger A. Pielke Jr. and others, "Normalized Hurricane Damage in the United States: 1900–2005," Natural Hazards Review, vol. 9, no. 1 (February 2008), www.nhc.noaa.gov/pdf/NormalizedHurricane2008.pdf (1.75 MB); and William D. Nordhaus, "The Economics of Hurricanes and Implications of Global Warming," Climate Change Economics, vol. 1, no. 1 (May 2010), www.worldscientific.com/doi/abs/10.1 142/S2010 007810000054.

[34] See Robert Mendelsohn and others, "The Impact of Climate Change on Global Tropical Cyclone Damage," Nature Climate Change, vol. 2 (2012), pp. 205–209, www.nature.com/nclimate/journal/v2/n3/abs/nclimate1357.html.

[35] See Laura A. Bakkensen and Robert O. Mendelsohn, "Risk and Adaptation: Evidence From Global Tropical Cyclone Damages and Fatalities," Journal of the Association of Environmental and Resource Economics (forthcoming).

[36] One study concludes that the global cost of flooding caused by rising sea levels is likely to be greater than the cost of building levees to avoid such damage. Further, the authors find that flood damage by the end of the century is more sensitive to the applied protection strategy than it is to variations in climate and socioeconomic scenarios. See Jochen Hinkel and others, "Coastal Flood Damage and Adaptation Costs Under 21st Century Sea-Level Rise," in Proceedings of the National Academy of Sciences, vol. 111, no. 9 (March 2014), www.pnas.org/content/111/9/3292.abstract. Even when measures are cost-effective, some researchers suggest that they may not be undertaken because they address low-level risks that may be overshadowed by other issues. Moreover, some researchers suggest that adaptive measures, such as the construction of seawalls, may inadvertently increase damage from extreme events. For example, by offering protection from less damaging (but more frequent) hurricanes they may encourage development and thus increase damage from less frequent but stronger Category 4 and 5 storms. (For a discussion of those points, see a literature review by Carolyn Kousky, "Informing Climate Adaptation: A Review of the Economic Costs of Natural Disasters," Energy Economics, vol. 46 (November 2014), pp. 576–592, www.sciencedirect.com/science/article/pii/S0140988313002247.

[37] Based on personal communication with Laura Bakkensen.

In: Hurricane Damage
Editor: Martha Atkinson

ISBN: 978-1-53610-035-8

Chapter 3

THE NATIONAL FLOOD INSURANCE PROGRAM: FACTORS AFFECTING ACTUARIAL SOUNDNESS*

Perry Beider

ABSTRACT

The vast majority of homes and small commercial buildings that are insured against flood damage in the United States are covered by the National Flood Insurance Program (NFIP), which is administered by the Federal Emergency Management Agency (FEMA) in the Department of Homeland Security. Although the flood insurance program had been largely self-sustaining in the past, it had to borrow about $17 billion from the federal Treasury to pay claims after the catastrophic hurricanes of 2005. That borrowing has highlighted questions about the program's financial health, including the actuarial soundness of the premium rates charged on policies that are not explicitly subsidized and the cost of paying claims for properties that have suffered multiple flood losses.

This Congressional Budget Office (CBO) paper—prepared at the request of the Ranking Member of the Senate Committee on Banking, Housing, and Urban Affairs—explains how FEMA sets "full-risk" (actuarially based) premium rates for the flood insurance program. The paper then discusses various reasons for concern that those rates may not

* This is an edited, reformatted and augmented version of a report issued by the Congressional Budget Office, November 2009.

be adequate to cover the total expected costs associated with the program's full-risk policies. The report also addresses other aspects of the NFIP, including the impact of insured properties that have flooded more than once and the U.S. market for flood insurance from private companies. In keeping with CBO's mandate to provide objective, impartial analysis, this report makes no recommendations.

INTRODUCTION AND SUMMARY

The main source of insurance against flood damage in the United States is the National Flood Insurance Program (NFIP), administered by the Federal Emergency Management Agency (FEMA). Some 21,000 communities, covering about 98 percent of the U.S. population, participate in the program. Almost all buildings in participating communities are eligible to be insured through the NFIP, except for structures that are built entirely over water or largely below ground. Certain buildings, such as those with a mortgage from a federally regulated lender, must be insured if they are located in areas at high risk of flooding, though compliance with that requirement is incomplete. In all, the NFIP had 5.6 million policies in force as of July 31, 2009, with a total insured value of $1.2 trillion and total annual premiums of $3.1 billion.[1]

The program encompasses three main activities:

- Offering insurance for flood risks—coverage is available in amounts up to $350,000 for residential properties ($250,000 for the structure and $100,000 for the contents) and up to $1 million for nonresidential properties ($500,000 each for the structure and contents);
- Mapping flood hazards—for each participating community, FEMA creates a flood insurance rate map that documents areas at risk of flooding; and
- Setting minimum requirements for building codes and floodplain management practices in participating communities.

Together, those components of the NFIP address the interrelated goals of reducing exposure to flood damage, allowing floodplains to play their natural beneficial roles, making it easier for flood victims to recover when damage occurs, and reducing federal costs (by charging premiums for flood insurance rather than providing disaster aid after a flood at no charge to the recipients).

Most NFIP policies are issued at "full-risk" premium rates that FEMA considers actuarially sound—that is, sufficient on average to cover the total flood claims and administrative costs for those policies based on the agency's maps and its estimates of the frequency of different size floods. However, for more than 1 million policies—about one-fifth of the total—premium rates are explicitly subsidized. Those policies mainly cover older structures in areas at high risk of flooding.

The subsidized policies give the NFIP a built-in actuarial deficit, which the Congressional Budget Office (CBO) currently estimates at about $1.3 billion per year. That actuarial deficit was somewhat obscured for many years by the program's operating results. In 2005, however, the NFIP experienced an unprecedented volume of claims resulting from Hurricanes Katrina, Rita, and Wilma. Total payments on those claims were greater than the total for all of the program's previous years combined and led the NFIP to borrow about $17 billion from the Treasury, an amount roughly equal to the program's net loss on insurance operations for that year (see Figure 1).

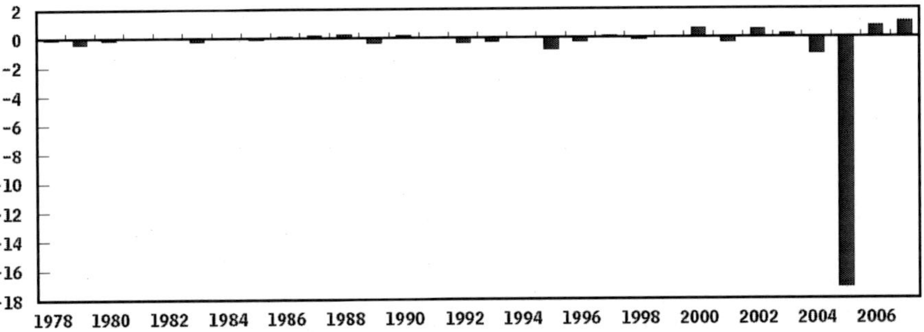

Source: Congressional Budget Office based on Thomas L. Hayes and D. Andrew Neal, *Actuarial Rate Review: In Support of the Recommended May 1, 2009, Rate and Rule Changes* (Federal Emergency Management Agency, National Flood Insurance Program, August 2009), Exhibit B1, available at www.fema.gov/library/viewRecord.do?id=3742.

Notes: NFIP = National Flood Insurance Program.

These data exclude the federal policy fee paid by policyholders and the expenses covered by the fee. The data also exclude about $1.5 billion in payments for principal and interest on the program's debt since 2005.

Figure 1. The NFIP's Net Profit or Loss on Insurance Operations, 1978 to 2007; (Billions of nominal dollars).

The 2005 loss highlighted a number of factual and policy questions about the NFIP's financial position. This CBO paper addresses two sets of factual questions.

First, does evidence suggest that the program's full-risk premium rates are actuarially sound, or does it suggest that those rates implicitly carry hidden subsidies from taxpayers?

Historically, the NFIP's full-risk premiums have been too low to cover the flood claims and administrative costs of the policies insured at those rates. Even before the 2005 hurricanes, the program's total claims and expenses on full-risk policies since 1978 exceeded its total premium income from those policies by about 5 percent. Taking into account the large 2005 losses, cumulative costs on those policies are almost twice the size of total income from the policies. That experience does not directly imply that current premium rates are too low, however, because rates have risen over time and because the frequency of future catastrophic years like 2005 is highly uncertain. In part because of that uncertainty, CBO does not have enough information about the current distribution of flood risks to calculate whether the present full-risk premiums are actuarially adequate.

Analyzing the methods that FEMA uses to set the full-risk rates does not yield definitive answers either. Some aspects of those methods tend to contribute to an actuarial surplus—the primary one being the additional 10 percent that FEMA includes in the rates in high-risk areas (20 percent in some high-risk coastal areas) as a safety margin for uncertainty. Other aspects of the agency's rate-setting methods tend to contribute to an actuarial deficit. FEMA is not reviewing its flood maps every five years as required by law, and some older maps do not reflect significant changes in local conditions that tend to increase the risk of flooding, such as coastal erosion, increases in sea level, land development, and reductions in the capacity of river channels. In addition, evidence suggests that climate change has increased the risk of flooding from rivers and perhaps also from coastal storms, making FEMA's models of flood frequencies out of date. Those issues may warrant attention regardless of the overall adequacy of the program's full-risk rates.

Second, to what extent are the NFIP's losses attributable to properties that have experienced multiple floods? How is FEMA responding to the risks posed by such properties?

Currently insured repetitive-loss properties (RLPs)— defined by FEMA as properties that have been the subject of two or more flood-claims payments greater than $1,000 apiece in any 10-year period—account for 2 percent of current policies and 3 percent of current premiums but about 12 percent of total claims since 1978. More than half of the roughly 71,000 RLPs that are now insured have had only two losses leading to such payments, and their experience may reflect random chance rather than truly above-average risks. However, about 23,000 RLPs nationwide have been the subject of at least four claims payments while insured, and 10,000 of those properties have prompted six or more payments.

FEMA's approach to reducing the cost of repetitive-loss properties focuses more on taking steps to mitigate the worst flood risks—such as elevating, relocating, flood-proofing, or demolishing properties—than on charging higher premiums for flood insurance. Indeed, more than half of the policies covering RLPs in high-risk areas have rates that are explicitly subsidized. Policies covering RLPs in low- to moderate-risk areas do have higher average premiums than non-RLPs in those areas (because they are ineligible for the lower-priced "preferred-risk policy" rates that FEMA charges on most policies in those areas). Even those higher average rates, however, are not based directly on the claims experience of individual properties or of repetitive-loss properties as a whole.

ACTUARIALLY SOUND VERSUS SUBSIDIZED PREMIUM RATES

When lawmakers created the NFIP, they faced a trade-off in deciding whether the premiums charged for flood insurance should be actuarially sound—that is, high enough to cover the full expected value (the average of all possible outcomes, weighted by their probabilities) of flood claims and administrative expenses for the insured properties—or subsidized. Premiums that were actuarially sound would be unattractively high for many properties that were built before the extent of their flood risk was documented, especially for properties that face repeated flooding. High premiums for such properties might discourage communities from joining the NFIP. Conversely, subsidized premiums would produce a cumulative deficit for the NFIP, meaning that the program would not be self-sustaining. Lower premiums could encourage more property owners to buy flood insurance and thus help reduce federal spending on disaster aid after floods. But total federal costs related to floods could increase if the NFIP's cumulative deficit exceeded the savings on disaster aid. (For more information about other federal programs related to flood prevention or recovery, see Box 1.)

Subsidies in the NFIP

The decision that lawmakers incorporated into the NFIP's founding legislation, the National Flood Insurance Act of 1968, was to subsidize the premiums charged for policies covering certain structures in high-risk flood areas. (Subsidized rates also apply to policies covering only the contents of those structures.) Program designers expected that the subsidized structures would be replaced over time as they became damaged or aged past their useful life. The proportion of policies that are subsidized has indeed gradually declined, but it is still about one-fifth.[2] More than 90 percent of those policies are for structures that were built before their community's flood insurance rate map (FIRM) was created or before 1975, whichever is later.[3] Such properties remain eligible for the so-called pre-FIRM subsidy if they are bought or sold, but they lose that eligibility if they experience flood damage worth at least 50 percent of their value or if improvements are made that increase their value by 50 percent or more.

Box 1. Other Flood-Related Federal Programs

The National Flood Insurance Program (NFIP) is one of many federal programs, both within and outside the Federal Emergency Management Agency (FEMA), that focus on preventing floods or helping people recover from flood damage. FEMA administers several grant programs that offer funding to state, local, and tribal governments for planning and carrying out flood-mitigation projects.[1]

Two of the FEMA programs apply to all types of natural disasters:

- The Hazard Mitigation Grant Program provides funding to states that have experienced an event that the President has officially declared a major disaster. The amount of funding is determined as a percentage of the total sum allocated for public and individual assistance after the disaster.
- The Pre-Disaster Mitigation Program awards funds primarily on the basis of a nationwide competition for proposed projects and planning activities.

Three other FEMA programs focus specifically on mitigating flood risks:

- The Flood Mitigation Assistance Program allocates funding to states on the basis of how many NFIP policies and repetitive-loss properties they have.
- The Repetitive Flood Claims Grant Program focuses on states and communities that do not participate in the Flood Mitigation Assistance Program because they lack the required matching funds or the capacity to manage the activities.
- The Severe Repetitive Loss Program allocates funds mainly on the basis of the number of severe repetitive-loss properties in each state.[2]

FEMA also provides recovery aid to public and private entities affected by Presidentially declared disasters. Funds for individuals—which are currently capped at $30,300 per household (an amount that is adjusted annually for inflation)—may be used for things such as temporary housing, home repair, moving and storage costs, and medical expenses. However, most federal disaster assistance to individuals and businesses takes the form of low-interest loans from the Small Business Administration (SBA). The loans are available in the event of a disaster declaration by the President or the SBA. The availability of both FEMA grants and SBA loans depends on Congressional appropriations.

From the point of view of flood victims, NFIP coverage has three advantages over federal disaster aid. It is available in larger amounts than the individual assistance from FEMA; it covers events that are not the subject of a disaster declaration; and unlike SBA loans, flood-claims payments do not have to be repaid. However, NFIP coverage requires premium payments, and benefits are limited by the terms of the insurance policy.

Flood insurance and disaster assistance both serve the basic purpose of helping flood victims recover. Thus, the two approaches substitute for each other to some extent—indeed, one of the rationales for the NFIP is that it helps reduce the demand for disaster assistance. Conversely, the expectation that aid will be provided in the event of a flood (whether by the government or by other entities, such as the Red Cross) may decrease the demand for NFIP coverage. That effect is an example of the broader problem that disaster assistance lessens the economic incentives to reduce one's exposure to disaster losses—in other words, it subsidizes risky behavior. Various features of the NFIP attempt to limit the adverse effects of that subsidy: Participating communities must meet the program's minimum requirements for floodplain management and building codes, and the owner of a property in a high-risk flood zone who receives federal disaster aid after a flood must buy and maintain NFIP coverage to remain eligible for assistance after a future flood.

Besides FEMA and the SBA, other federal agencies involved in flood prevention or recovery include the Army Corps of Engineers, which builds levees and other flood-control structures, and the Department of Agriculture, whose Federal Crop Insurance Program covers farmers against losses from floods, pest infestations, droughts, and other natural disasters.

[1] For an overview, see Federal Emergency Management Agency, Mitigation Grant Programs Fact Sheet (April 2009), available at www.fema.gov/library/view Record.do?id=2065, and FY 2010 Hazard Mitigation Assistance Unified Guidance (June 2009), available at www.fema.gov/library/viewRecord. do?id=3649. The effectiveness of projects funded by FEMA to mitigate risks from floods, earthquakes, or windstorms (including hurricanes) is discussed in Congressional Budget Office, Potential Cost Savings from the Pre-Disaster Mitigation Program (September 2007).

[2] Repetitive-loss properties (RLPs) are ones that have been the subject of at least two flood-claims payments exceeding $1,000 apiece in any 10-year period. The Severe Repetitive Loss Program focuses on the subset of RLPs that are residential properties and have prompted at least four claims payments of more than $5,000 each, or two or more payments that together exceed the property's value, with at least two of the payments occurring within 10 years of each other.

On average, the premiums charged for subsidized policies are about 35 percent to 40 percent of their full-risk level, FEMA estimates.[4] The actuarial shortfall that results from those subsidies (currently about $1.3 billion a year) has drawn greater scrutiny following the program's unprecedented losses in 2005. Previously, the NFIP had been able to cover its costs, borrowing relatively small amounts from the Treasury on a few occasions to pay claims and then quickly repaying the loans with interest. But the $17 billion it borrowed from the Treasury in response to the surge in claims from the 2005 hurricanes is more than five times the program's annual income from premiums. The NFIP can pay interest or even repay principal on that debt when it experiences a year with low claims; indeed, it has made $2.9 billion in interest and principal payments since 2005. Nonetheless, it is very unlikely to be able to retire the debt without substantially raising premiums. The debt grew to $19 billion by the end of fiscal year 2009, and given the actuarial deficit built into the NFIP, that figure can be expected to continue growing over the long term. Even if the program managed to generate large surpluses during an extensive run of low-claims years, using those surpluses to retire the debt instead of accumulating them as reserves would make it more likely that the program would have to borrow again in the future.

Subsidized rates for flood insurance have implications not only for the federal budget but also for the expected amount of flood losses. The program's explicit subsidies lower the cost of living in high-risk properties. Moreover, they undermine the incentives for policyholders to carry out mitigation measures—such as flood-proofing or elevating properties—because the subsidized rates are not affected by such measures.[5] The explicit subsidies do not encourage development in risky areas because they are not available for new construction. However, any implicit (that is, unacknowledged) subsidies that may exist in the full-risk rates if FEMA underestimates the actual risk of flooding may lead to excessive development in risky areas. (The same may be true for the expectation that disaster assistance will be available after a major flood; see Box 1.)

Besides explicit and possibly implicit subsidies from taxpayers, a third type of subsidy in the NFIP is the "cross-subsidy" to some policyholders from others who are charged premiums above their full-risk rates (see Appendix B). If cross-subsidies completely balance out, they do not create financial risk for taxpayers. But like the other types of subsidies, cross-subsidies distort the perceived riskiness of the properties involved. Moreover, the rates that are raised to provide cross-subsidies may discourage some potential policyholders from buying coverage.

The Meaning of "Actuarial Soundness"

Actuarial soundness and actuarial adequacy are common concepts in the insurance industry, but their specific definition can vary depending on the context. In this report, both terms refer to the ability of premiums to cover the expected value of flood claims and administrative costs, and the NFIP is considered subsidized by taxpayers to the extent that premiums fall short of that amount. Those definitions do not account for any economic value associated with the risk that is transferred from policyholders to taxpayers, but they are consistent with the current cash-based treatment of the NFIP in the federal budget.[6]

The definition of actuarial soundness used here has three noteworthy implications:

- The NFIP must cover not only costs directly related to insurance policies and claims but also costs for flood insurance studies, floodplain management, and administrative personnel. The Omnibus Budget Reconciliation Act of 1990 directed FEMA to charge policyholders a fee to cover those and other implementation costs of the program.[7] Currently, that fee is $35 per year for most policyholders and yields about $140 million annually.[8] The rule that policyholders be charged for implementation costs has one current exception: The Map Modernization program, a multiyear effort to convert FEMA's flood maps to digital form, is funded mainly from separate appropriations. This CBO paper focuses on FEMA's methods for estimating the costs of claims; it does not explore the accuracy of FEMA's estimates for other administrative or program costs.
- Proposals to increase the amounts or types of coverage available under the NFIP (for example, by raising policy limits or offering coverage for wind damage) and pricing the additional coverage to recover its full costs would not improve the program's actuarial soundness as defined here. Successfully implementing such proposals would reduce the program's actuarial deficit as a percentage of its premium income—a factor that would be relevant to a private insurer because it could reduce the risk of bankruptcy. But doing so would not directly decrease the NFIP's actuarial deficit as measured in dollars.[9] That deficit can be decreased in only two ways: by reducing the extent to which some coverage is sold at premiums below the full-risk level or by increasing the extent to which some coverage is sold at premiums above the full-risk level (that is, by increasing cross-subsidies from some policyholders to others).
- Rates that were actuarially sound as defined here would not be high enough, in general, to allow private insurers to compete with the NFIP. Unlike the NFIP's rates, actuarially sound premiums in the private sector need to reflect the cost of capital—that is, to compensate investors for the use of their funds and the risk they are accepting. Nonetheless, some flood insurance in the United States is provided by private insurers, generally in ways that complement rather than substitute for coverage offered under the NFIP (see Box 2).

AN OVERVIEW OF THE NFIP'S FULL-RISK RATES

In setting full-risk (or actuarially based) premium rates for flood insurance, FEMA groups together properties in geographic zones based on the risk of flooding. For each zone, the agency estimates the expected annual flood losses on an insured property as a percentage of the property's value, using data (where available) on floodwater flows, water heights, and damage from water at that height and taking into account estimated probabilities of very large floods that have not yet occurred. FEMA then determines premiums on the basis of those expected losses, after adjusting for deductibles and other factors. Between 1978 and 2004, the total amount of premiums that the NFIP collected on full-risk policies roughly equaled the total amount of claims it paid. But claims from hurricanes in 2005 have left the program with a large debt.

Geographic Zones

The methods that FEMA uses to calculate full-risk rates vary by zones—areas shown on the agency's flood insurance rate maps that are defined by the degree or type of flood risk and the amount of information available.

Box 2. Flood Insurance from Private Companies

Private firms provide relatively little flood insurance in the United States, and information about private flood insurance is spotty. What is known, however, is that private companies rarely compete directly with the National Flood Insurance Program (NFIP). Most private policies address special needs that the NFIP does not serve as well or at all.

One market for private flood insurance consists of property owners who need coverage above the amounts available in the NFIP (a maximum of $250,000 for a residential structure, $100,000 for its contents, and $500,000 each for a nonresidential structure and its contents). For example, private insurers dominate the market for flood coverage on large commercial properties (excluding those that are self-insured). Data on that coverage are not readily available, in part because such insurance is typically provided as part of some broader coverage—either overall commercial insurance, often using the NFIP limits as the deductible, or "difference in conditions" coverage that insures against losses from a variety of disasters. For residential properties, a 2008 survey by an industry consulting firm indicates that in that year, 13 private sources offered "excess-layer" policies to insure residential properties against flood damage above the amount covered by a primary policy (whether from the NFIP or elsewhere). Those private sources consisted of seven companies that issue the policies they sell and six brokerage firms that represent other companies, mostly Lloyd's of London.[1]

Primary flood coverage for residential properties was available from three companies and three brokerages, according to the 2008 survey. The companies selling their own flood policies focused on clients with high net worth for whom they were already writing homeowner's insurance; such coverage was offered only for homes in areas at low-to-moderate risk of flooding. Some of the policies sold by brokers were for properties on coastal barrier islands where NFIP insurance is prohibited under the Coastal Barrier Resource Act or in communities that did not participate in the NFIP. Other policies were sold in conjunction with excess-layer coverage on high-value homes.

The best-studied segment of the U.S. market for private flood insurance is the market for policies that mortgage lenders buy on behalf of homeowners who are required to carry flood insurance but have not purchased it themselves. That requirement applies to any property in a 100-year floodplain that carries a mortgage from a federally regulated lender or a mortgage that is insured or purchased by Fannie Mae, Freddie Mac, or another government-sponsored enterprise or federal agency. Although lenders can choose to purchase such coverage from the NFIP, they rarely do so, according to a report by the RAND Corporation. Instead, they typically buy private insurance policies that have certain features tailored to meet the lenders' needs: For example, policies purchased from private insurers take effect on the date the lender sends a letter to a homeowner noting the requirement for flood insurance coverage.[2] Using data from various sources for different periods from mid-2004 to early 2006, the RAND study estimated that private insurers had roughly 130,000 to 190,000 lender-placed residential policies in effect at any one time, compared with the NFIP's 5 million or so residential policies. The number of lender-placed residential policies written during a year may be much higher than the above range, however, because few of those policies remain in effect for a full year. Property owners, to whom the costs of the policies are passed on, typically replace them with less expensive coverage from the NFIP.[3]

The fact that the market for private flood insurance is small does not indicate that the NFIP's full-risk premium rates are too low to cover the flood risks and associated administrative costs. Private insurers must set their rates high enough to cover their cost of capital (including reserves for losses incurred but not yet reported), and that cost would be influenced by the high variability of flood losses. Thus, the NFIP's rates could be actuarially sound as the term is used here and still be too low for private insurers to compete with them.

[1] Bruce A. Bender, "Alternatives to the NFIP" (unpublished table, Bender Consulting Services, May 2008). Homeowners typically have fewer than 13 private companies to choose from, however, and some have none at all. Some of the companies do not offer flood coverage in all states, and most of them are selective about the properties they insure, rejecting those that face the highest risks.

[2] A lender can also buy an automatic coverage endorsement that allows the insurance to take effect even earlier—for example, on the date the lender refers the property to a tracking firm for verification of its flood insurance status; Lloyd Dixon and others, *The Lender-Placed Flood Insurance Market for Residential Properties* (Santa Monica, Calif.: RAND Cor poration, 2007), pp. 25–28.

[3] Ibid., pp. xiv, xv, 31.

The labels used on the FIRMs have changed over time. Today, the major categories are A zones, V zones, and Zone X:

- A zones include all 100-year floodplains—that is, areas where the risk of flooding is thought to be at least 1 percent annually—with the exception of areas designated as V zones.
- V (for "velocity wave action") zones are coastal areas in 100-year floodplains where waves add at least three feet to the water level that would be reached during a 100-year flood in the absence of wave action.[10]
- Zone X includes essentially all areas outside 100-year floodplains. (The exceptions are areas where no analysis of flood hazards has been conducted. Such areas, which are labeled Zone D, are rare and account for very few properties insured by the NFIP.)

FEMA divides the A and V zones into a variety of subcategories, which are described in Appendix A.

Properties covered by the NFIP are concentrated in coastal areas. In a sample of 20,000 properties that FEMA analyzed for CBO (including those insured at both subsidized and actuarially based rates), about 40 percent were in zip codes containing at least one V-zone property. However, only 1 percent were in the V zones; the rest were far enough from the shore not to be susceptible to damage from three-foot waves during a 100-year flood.[11] Thus, the A zones and Zone X contain nearly all of the properties insured at full-risk rates (the A zones currently account for half of full-risk policies, and Zone X accounts for just under half). Even if coastal properties lie outside the V zones, they may face higher flood risks, on average, than their inland counterparts, for reasons discussed below.

Estimating Expected Losses

Expected losses for properties in 100-year floodplains (A and V zones) are generally calculated using three types of information:

- Local data on the frequency of floodwater flows of different sizes;
- Representative topographic profiles to convert floodwater flows to water heights (which are greater, for any given volume of water, in a narrow valley than on a broad plain); and
- Damage functions that specify the percentage damage a given depth of flooding does to a particular type of structure. (The functions differ by zone type because properties in V zones are subject to wave action and damage to piers and pilings below the lowest floor.)

Analyzing local data on floodwater flows is complicated by the relatively short historical records of floods that exist in some inland areas. FEMA has concluded that when its analytic methods use data with short records, it underestimates flood risks. All else being equal, that factor tends to make the mapped boundaries of 100-year floodplains too small and to understate the risks within the mapped floodplains. To compensate for the latter bias, FEMA includes an adjustment in calculating the A-zone rates. However, the adjustment factors have not changed since the 1980s, although additional data have been collected and maps have been updated, and they are now likely to be incorrect. (For details about FEMA's correction for short historical records, see Appendix A.)

FEMA uses the three types of information listed above to produce tables for properties in A and V zones that show expected percentage losses as a function of a structure's type and of its elevation relative to the 100-year flood (known as its base flood elevation). The tables may be revised annually as FEMA updates the damage functions to reflect its most recent experience with claims.

For properties outside 100-year floodplains (in Zone X), two different sets of full-risk premium rates are used. Roughly one-fourth of the structures insured at Zone X rates are covered under so-called standard rates and three-fourths under the rates for preferred-risk policies. Preferred rates are available for a property in Zone X that is not grandfathered (see below) and whose past and present owners have received no more than two payments of NFIP claims or federal disaster aid—and no more than one payment exceeding $1,000—for floods that damaged the property. Specific combinations of structure and contents coverage and specified amounts of contents-only coverage are available, and the premiums are determined by the professional judgment of FEMA's actuaries on the basis of claims experience.

For the Zone X properties insured at standard rates, FEMA uses estimates of expected annual losses to determine premiums, as it does for properties in the A and V zones. However, the agency has no set of estimated probabilities of floods of all sizes for Zone X to use in estimating expected annual losses. Instead, it extrapolates the expected losses for properties insured at standard rates from the average losses observed to date on those properties by assuming that the ratio of expected to observed losses is the same as in the main A zone. FEMA's analysts make that calculation with and without the losses from 2005 and use their professional judgment to combine the two results. (The agency has commissioned a study to provide recommendations on the appropriate treatment of the 2005 experience.)

Roughly 20 percent to 25 percent of Zone X standard policies cover properties that are actually located in an A or V zone—that is, within a 100-year floodplain. Those properties were grandfathered at the Zone X standard rate by FEMA after they were remapped from Zone X into a high-risk zone. The agency allows such grand-fathering partly to hasten the incorporation of new information into communities' floodplain management efforts by reducing property owners' opposition to the revised flood maps. To the extent that flood losses on the grand-fathered properties are higher than those on "true" Zone X properties paying the standard rate, the grandfathered properties are cross-subsidized by the other standard ratepayers, because the higher losses on the grandfathered properties contribute to the observed damages on which FEMA bases its standard rates. Such grandfathering may not undermine the actuarial soundness of the NFIP as a whole. But the burden of funding the cross-subsidy may discourage owners of some Zone X properties that are ineligible for preferred-risk rates from buying or maintaining coverage. (That effect grows larger as the percentage of grandfathered properties increases.)

Translating Expected Losses into Premium Rates

Once it has estimated the expected losses for properties of various types and elevations in each flood zone, FEMA converts those estimates into premium rates by using certain adjustment factors. Four of its general adjustments are noteworthy:

- All premiums in 100-year floodplains are increased by a contingency-loading factor of 10 percent in the A zones and 20 percent in the V zones. Contingency loads are commonly used in the insurance industry to provide a safety margin for uncertainty—such as the uncertainty about the assumed frequencies of catastrophic floods.[12] FEMA's current contingency loads were increased from 5 percent in the A zones and 10 percent in the V zones; they remain under review.
- The Community Rating System (CRS) gives discounts to policyholders in communities that take additional steps to reduce flood hazards. In implementing that system, FEMA scales up its base (undiscounted) rates by a factor calculated to offset the discounts and thus keep the CRS program as a whole revenue neutral.[13] Tailoring the rates more specifically to a community's level of risk should help reduce cross-subsidies among communities. Even if the CRS creates new cross-subsidies because some of the discounts overstate the effectiveness of the credited activities, it probably has little effect on the overall actuarial soundness of the flood insurance program.[14]
- As needed, FEMA reduces the rates it would otherwise charge to avoid annual increases of more than 10 percent. The 10 percent cap is specified in law, and FEMA staff interpret it to mean that the premiums for a given combination of flood zone and structure type should not rise by more than that amount. FEMA also interprets the

cap as indicating that annual increases in specific rates—such as the rate for single-family residences in the main V zone that lie one foot above the base flood elevation and are insured to 60 percent of their value—should not exceed 10 percent by more than a few percentage points.

- As necessary, FEMA also raises the rates it would otherwise charge to ensure that it covers its estimated fixed costs on each policy. For instance, the agency's published rates for coverage of a single-family residence do not go below 24 cents per $100 for the basic layer of coverage (up to $50,000) or below 8 cents per $100 for coverage above that layer.

Other adjustment factors that FEMA uses in converting estimated losses into premium rates—such as for deductibles, loss-adjustment expenses, and underinsurance— reflect standard practices in the insurance industry.[15] (More specifics about FEMA's rate-setting methods are provided in Appendix A.)

In most recent years, FEMA has made an additional adjustment in the rates for full-risk properties in coastal zones, scaling up the rates determined by the above methods as much as possible within the constraints of the 10 percent cap.[16] Those increases are FEMA's response to a combination of two factors: coastal erosion and the cap itself. As discussed in more detail below, erosion increases flood depths in coastal areas relative to the depths shown on flood maps. Thus, policies on structures in an area subject to coastal erosion tend to become increasingly underpriced as time passes since the last remapping. Moreover, because development tends to be denser the farther one goes from the shoreline, the number of properties affected by such underpricing increases over time as erosion continues to claim the less developed shore land.

FEMA staff expect the erosion problem to grow to the point where the premiums charged in V zones would have to rise much more than 10 percent per year to keep pace. Because the cap will preclude such increases if it remains in force, FEMA is raising current rates above the levels it believes are necessary to cover today's risks in order to prefund future risks and allow for higher rates in the future.

Table 1. The NFIP's Experience with Full-Risk Policies, by Risk Zone (Billions of dollars)

	A Zones	V Zones	Zone X	Total
	1978–2006 (Full period of data)			
Number of Policy-Years	28,710,924	442,585	24,834,155	53,987,664
Number of Claims	225,072	7,225	263,574	495,871
Gross Premiums	6.5	0.5	5.6	12.6
Gross Claims and Expenses	13.8	0.5	9.9	24.2
Net Profit	-7.3	*	-4.3	-11.6
Net Adjusted Profit[a]	-6.9	0.2	-5.6	-12.3
	1978–2004 (Omitting catastrophic losses of 2005)			
Number of Policy-Years	24,565,935	362,880	21,727,759	46,656,574
Number of Claims	138,736	5,149	215,547	359,432
Gross Premiums	5.2	0.3	4.6	10.2
Gross Claims and Expenses	4.8	0.3	5.6	10.7
Net Profit	0.4	*	-0.9	-0.5
Net Adjusted Profit[a]	1.3	0.2	-2.1	-0.5

Source: Congressional Budget Office based on data from the Federal Emergency Management Agency. Notes: NFIP = National Flood Insurance Program; * = between -$50 million and $50 million.

V zones are coastal areas subject to waves of three feet or more during a 100-year flood; A zones consist of all other coastal or inland areas within 100-year floodplains; and Zone X comprises areas outside 100-year floodplains.

[a] Net adjusted profit is the cumulative profit from past years adjusted for inflation and for changes in the number of policies and the mix of policies by zone, but not for changes in premium rates (other than inflation). Net adjusted profit represents the cumulative profit or loss in real dollars that the NFIP would have experienced if it had insured the current number and mix of policies each year at historical premium rates (adjusted for inflation). Adjusted profit does not provide a direct measure of the soundness of current rates, however. Some past rates were significantly lower than current rates, especially in the early years of the period.

Premiums versus Claims over Time

Historically, premium income from the NFIP's full-risk policies fell short of the total costs of those policies even before the catastrophic losses of 2005. Between 1978 and 2004, premiums for such policies totaled $10.2 billion in nominal dollars, while claims and expenses totaled $10.7 billion (see Table 1). The result was a loss of $0.5 billion over that period on policies with full-risk rates. Adding in data for 2005 and 2006 brings total premiums to $12.6 billion and total claims and expenses to $24.2 billion. Thus, with the large losses of 2005 factored in, the NFIP ran a loss on full-risk policies over the whole 1978–2006 period of $11.6 billion in nominal dollars or $12.3 billion adjusted for inflation and for changes in the number and mix of policies.[17] The latter measure corrects for the effects of general price inflation and program growth that would otherwise make later years more important than earlier years to the NFIP's actuarial performance during that period.

On the surface, the program's loss over the 1978–2006 period might suggest that the full-risk premiums were only about half as large as needed. However, those figures cannot be taken at face value in evaluating the adequacy of the actuarially based rates because flood losses on the scale seen in 2005 are expected to recur much less often than once in 29 years. (The true "return interval" is highly uncertain, but future catastrophes on that scale are far from impossible: A sufficiently powerful storm striking Miami or Houston could lead to even larger losses for the NFIP).

Moreover, the profit figures shown in Table 1 are backward-looking and thus do not provide direct evidence of an actuarial shortfall in current rates. Specifically, the figures are not adjusted for changes in premium rates other than inflation, so they reflect FEMA's experience with the rates it charged (in nominal or real dollars) over the years, not the results it would have gotten with current rates. Many premiums were much lower in the first several years of the period than they are today: roughly 50 percent lower for the A zones, on average, and one-third to two-thirds lower for Zone X (although that may reflect differences in the amount of coverage as well as in the premium rates). With the early years of that period omitted, the NFIP ran an adjusted net profit of $0.6 billion on its full-risk policies over the two decades from 1985 to 2004.

Although Table 1 cannot settle the question of the actuarial adequacy of the program's full-risk premiums, it does provide some useful insights. One notable fact is that more claims have been paid on fewer total policy-years in Zone X (outside 100-year floodplains) than in the A zones (within those floodplains). The rate of claims in the A zones is less than 1 per 100 policies per year, in part because of the building codes and floodplain management practices adopted by participating communities, which are designed to reduce the rate of flood damage. In contrast, the claims rate in Zone X is over 1 percent per year for the full 1978–2006 period and nearly that high without 2005 and 2006. One reason is that a small fraction of those policies cover properties that are actually inside a 100-year floodplain: They have been remapped into an A or V zone but remain classified as Zone X because they have been grandfathered to stay eligible for coverage at Zone X premiums, which are generally lower. The high claims rates for that zone also suggest that the floodplain boundaries on FEMA's maps may tend to be too small. That is particularly likely for floodplains in communities with short historical records of floods (see Appendix A).

The table also highlights the fact that, to date, the NFIP has not lost money on properties in V zones, even after the four hurricanes that struck Florida in 2004 and the unprecedented losses of 2005. That result reflects the narrow definition of V zones: Only a small share of properties in coastal areas are thought to be subject to storm waves of three feet or higher in a 100-year flood. Thus, the limited data available suggest that if the NFIP has any problem adequately pricing the risks of coastal flooding, the problem is concentrated not in the V zones but in the A and X zones in coastal areas.

ISSUES REGARDING THE SOUNDNESS OF THE NFIP'S FULL-RISK RATES

CBO cannot determine whether the full-risk premiums currently used in the flood insurance program are actuarially adequate on the whole. Some aspects of FEMA's rate-setting methods tend to promote an actuarial surplus on the full-risk policies, whereas others tend to produce a deficit. The historical data are insufficient to support a more definitive conclusion, particularly because estimates of the frequency of catastrophic floods are so uncertain and because flood risks may be changing over time. Over a period of 30 years, the full-risk premiums could be adequate or too high and still yield occasional deficits and even large cumulative debts after rare catastrophic floods. Conversely, those premiums could be too low and still yield surpluses in most years. One would need to analyze data spanning many decades (or even centuries, if the distribution of floods is changing over time) to distinguish between those two cases with confidence from the data alone.

Analysis of FEMA's rate-setting methods identifies some aspects that err on the side of creating an actuarial surplus. Of those, probably the most important are the contingency loads (safety margins) included in the premiums for properties in 100-year floodplains: 10 percent in the A zones and 20 percent in the V zones. The fact that V-zone rates are set higher than they would be otherwise in anticipation of future increases in erosion risk also tends to create a surplus in the short run—although FEMA expects those rates to be inadequate in the future when the rise in erosion risk outpaces the 10 percent annual limit on rate increases. A third possible source of surplus is overcompensation for bias in estimates of risks in areas with short historical records of floods (see Appendix A).[18]

Despite those factors, FEMA's full-risk rates may be too low overall because the agency uses some outdated or incomplete information in its rate-setting—specifically, in its maps of areas at risk of flooding and its models of flood frequency and severity. Many flood maps are too old to reflect recent erosion of coastlines, decreases in permeable ground area because of wetland loss and economic development, or increases in sea level. FEMA's policy of grandfathering properties in a previous zone or elevation class can also be seen as a case of basing rates on information that is out of date. Moreover, FEMA's models of flood frequency and severity may understate actual risks because they do not reflect the effects of global climate change. (Less important, the models also neglect the effects of storm-water flooding—flooding from water that has not yet entered a river, stream, or other channel—and the damage caused by debris carried by floodwaters.)

The factors mentioned above that tend to create an actuarial surplus in FEMA's full-risk rates offset the financial effect of the weaknesses in the flood maps and models to some unknown degree—partially, fully, or more than fully. Even in the unlikely case that the two sets of factors exactly balance each other out for the program as a whole, the effects on rates paid by individual policyholders will not all balance out. Rather, the result will be hidden cross-subsidies, with some policyholders paying more than the actuarially fair level for their coverage and others paying less. Other cross-subsidies in the NFIP are more easily identified (see Appendix B). In particular, properties grandfathered at Zone X standard rates and those in areas that are classified as Zone X because they are protected by levees or other flood-control structures (notwithstanding their additional risks from failure of those structures) are both cross-subsidized by other Zone X policyholders. In both cases, the higher risks faced by those properties contribute to the observed damages that FEMA uses in determining the Zone X standard rates for all properties. That full averaging of the higher and lower risks makes the net impact on the program's actuarial balance small or zero. (Grandfathered properties in the A and V zones are not directly cross-subsidized and do contribute to a shortfall, as explained at the end of the next section.)

Of course, actuarial soundness—that is, avoiding an implicit subsidy from taxpayers—is not the only goal set for the NFIP. The program is also intended to reduce flood losses and federal costs for disaster assistance and to facilitate recovery when floods occur. On those terms, the information gaps that lend a downward bias to the rates could be problematic even if the rates are not too low overall because of cross-subsidies. As noted above, rates that are below the actuarially fair level can lessen incentives for policyholders to reduce their flood risks (whereas rates above that level, serving as a source of cross-subsidies, may discourage some potential policyholders from buying coverage). At the same time, rates that are too low actuarially can also encourage communities to participate in the NFIP and thereby reduce federal disaster aid.

Outdated Flood Maps

In general, topographic information that is not up to date lends a downward bias to estimates of flood risks and thus contributes to the likelihood of an actuarial shortfall. By law, FEMA is required to review all of its flood maps on a five-year cycle; to date, it has not met that requirement, at least partly because it has lacked the resources to do so. Keeping the maps relatively current will be easier once FEMA completes its multiyear Map Modernization program, in which the agency is both updating maps and switching them from paper to digital form. The digitized maps can be made available over the Internet and are more precise and less costly to update.

How current are the maps used in the NFIP? Different estimates measure currentness in different ways. According to the Government Accountability Office (GAO), 50 percent of the program's roughly 106,000 maps were more than 15 years old in April 2008, and another 8 percent were 10 to 15 years old.[19] For its part, FEMA says that "68 percent of the nation's population has already received updated maps that meet, exceed, or at least approach the map quality targets of Map Modernization," and another 27 percent of the population has updated preliminary maps (those not yet approved by the local communities).[20] For purposes of actuarial soundness, the share of the population covered by up-to-date maps is more meaningful than the percentage of maps themselves, so GAO's figures may overstate the extent of the problem associated with outdated maps. However, FEMA's figures focusing on map quality do not address the concern that older maps fail to reflect changes over time in geographic factors that may tend to increase flood risks. Moreover, FEMA staff report that the majority of coastal maps are based on outdated analyses and some were drawn under an earlier set of specifications for analyzing the particular characteristics of coastal flood hazards.[21]

One factor that causes older maps to understate flood risks is shoreline erosion. When FEMA updates a map of an area subject to such erosion, the mapped flood depths take into account the erosion that could occur during a significant storm; accordingly, the insurance premiums are higher than they would be without the erosion risk. Once such a storm occurs and causes erosion, however, the mapped coastline and flood depths are no longer accurate. According to a 2000 study conducted for FEMA by the Heinz Center, losses of 50 to 100 feet of shoreline over 30 years (or 20 to 40 inches per year) are "fairly common" on the coasts of the Atlantic Ocean and the Gulf of Mexico.[22] The study estimated that without additional measures to protect coasts, about 25 percent of structures within 500 feet of the Atlantic, Pacific, Gulf, or Great Lakes shoreline and outside major cities—or about 87,000 structures in all, assuming no new development—will be lost to erosion within 60 years. At the projected rates of erosion, about 10,000 of those structures will be lost within 10 years. Using observed participation rates in the NFIP, the study estimated that costs to the program for erosion losses to insured properties would average about $80 million a year over 30 years (or about 5 percent of 1999 premiums), with costs rising over time.

For now, however, NFIP premiums may be high enough to cover the risks associated with the cumulative erosion expected between one map review and the next. FEMA staff have indicated that the agency does not have legal authority to base premiums on such erosion risks. Instead, FEMA has been raising rates on all properties in V zones as a rough substitute for targeting the specific areas with high erosion risks. Between 2000 and 2008, full-risk rates in the V zones rose by 78 percent.[23] (For more discussion of erosion risks in the NFIP, see Appendix C.)

Another factor that makes flood maps outdated is the loss of wetlands. In coastal areas, wetlands serve as natural buffers to storm surges, and in inland areas, they can reduce the extent of downstream flooding. As an example of their disappearance, coastal Louisiana loses an area of wetlands about the size of Manhattan each year, primarily because of levees, jetties, canals, and subsidence (land sinkage).

Even when it does not displace wetlands, ongoing development—particularly increases in the ground area covered by roads, buildings, and other impermeable surfaces—reduces the natural environment's ability to absorb or delay water flows during storms and floods. That issue is illustrated by the results of a study, commissioned by FEMA, that looked at selected river segments in four communities and examined the effects of planned development on the damage expected from a 100-year flood. The study found that planned development would increase expected damage to *current* buildings and their contents by about 20 percent in the studied area of Fort Collins, Colorado; by roughly 100 percent in parts of Mecklenburg County, North Carolina, and DuPage County, Illinois; and by more than 1,200 percent in the studied area of Harris County, Texas.[24] Those areas are not intended to represent the nation as a whole: They were selected for their potential for significant development. Moreover, the estimates assume that development occurs without any mitigation efforts, such as on-site retention of storm water (which is required in three of the four jurisdictions studied). Even so, the examples provide qualitative evidence that development can have an impact on the depth and coverage area of river flooding. The relevance of those case studies is supported by the view of representatives of the Association of State Floodplain Managers, who told CBO that the "vast majority" of FEMA's older maps understate current flood risks, in part because of increases in impermeable surface area.[25]

Yet another factor that affects the topographic information shown on FEMA's maps is climate change. As the global climate warms, sea level rises because of the thermal expansion of seawater and the loss of ice sheets in Greenland and Antarctica. The ongoing rise in sea level shortens the useful life span of flood maps of coastal areas. (Climate change may also affect the frequency and severity of floods. Those effects are discussed in the next section.) According to a working group of the Intergovernmental Panel on Climate Change, "There is strong evidence that global sea level gradually rose in the 20th century and is currently rising at an increased rate, after a period of little change between AD 0 and AD 1900."[26] From 1961 to 2003, sea level rose at an average rate of 1.8 millimeters (mm) per year. During the last 10 years of that period, from 1993 to 2003, the average rate was about 3.1 mm per year, although it is unclear whether that higher rate marks an increase in the long-term trend.[27] Even a small rise in sea level can move shorelines significantly inward, depending on the local terrain: In South Carolina's low country, for example, one vertical inch corresponds to roughly 200 horizontal feet. Besides moving shorelines, higher sea level also increases storm surges and erosion hazards.

FEMA staff acknowledge that outdated maps—particularly in areas subject to coastal erosion—pose a problem for the actuarial soundness of the flood insurance program.[28] The agency has created a five-year plan for flood mapping and public awareness of flood risks. Among its goals are ensuring that all populated coastal areas have updated flood-hazard data and that data for 80 percent of all flood hazards are new, updated, or still valid.[29] FEMA estimates that the plan's goals could be achieved with annual appropriations of $220 million through fiscal year 2014. FEMA staff are less concerned about maps' being outdated because of man-made changes in floodplains, such as new bridges and culverts, partly because federal regulations require communities that participate in the NFIP to report to FEMA on "physical changes affecting flooding conditions" no later than six months "after the date such information becomes available."[30] The agency produces an average of about 2,000 Letters of Map Revision per year to incorporate information on significant changes.

In the case of grandfathered properties, FEMA has up- to-date map information but chooses not to use it in setting premiums. FEMA staff say that most cases of grand-fathering probably involve properties that were outside the 100-year floodplain and have been remapped inside it but are still paying Zone X standard rates rather than an A-zone or V-zone rate. Some cases, however, involve A-zone properties that would have moved to a V zone, or A- or V-zone properties that would have remained in the same zone but been reclassified at a lower elevation. Assuming everything else is the same, those lower- elevation properties generally face higher flood risks than predicted by FEMA's models for the A and V zones. (The exceptions are cases where the lower elevation is still well above any level that floodwaters are likely to reach.) Because the "excess" flooding that those properties experience does not affect the models—and thus does not feed back into higher rates for all properties in those zones, as it does in Zone X—they may contribute to an actuarial shortfall.

Climate Change

Besides raising sea level, climate change could adversely affect the NFIP through such mechanisms as more severe rainstorms and increased hurricane activity. A June 2008 report on climate change that summarized and synthesized many previous scientific studies provided evidence of trends in rainfall that could increase losses from river flooding.[31] It also offered some evidence of adverse trends in hurricanes. FEMA has begun studying the implications of climate change for the NFIP and expects to complete its analysis in early 2010.[32]

Effects on Rainstorms

On the basis of several studies, the June 2008 report concluded that it is "highly likely" that rainfall from severe rainstorms has increased in recent decades. One of the studies found a 50 percent increase over the past century in the number of days with more than four inches of rain in the upper Midwest. Another of the studies found statistically significant upward trends in the amount of rain falling on the rainiest days (specifically, those in the top 0.3 percent of the distribution) in the South, Midwest, and upper Mississippi regions. The report's authors also analyzed 100 years of data on total precipitation over 90-day periods, which is relevant to efforts to predict the likelihood of major widespread (rather than localized) floods. They found that 90-day periods with total precipitation in the top 5 percent of the distribution occurred 20 percent more often in the past 25 years than in earlier 25-year periods.[33] Some scientists, however, argue that the evidence available to date does not show that climate change is affecting river flooding.[34]

Effects on Hurricanes

Researchers generally agree that hurricane activity has been greater in recent years than it was from 1970 through 1994 (when the number of hurricanes reaching the United States was low by historical standards) and that risks remain higher in the near term.[35] For example, one researcher estimates that the Power Dissipation Index for tropical storms—a measure that reflects the storms' frequency, intensity, and duration—tripled over the 30 years from the early 1970s to the early 2000s for hurricanes in the North Atlantic.[36] Globally, the percentage of storms in a five-year period rated as category 4 or 5 on the Saffir-Simpson scale more than doubled over roughly the same 30 years, from 16 percent in the 1970–1974 period to 36 percent in the 2000–2004 period.[37]

The key question for the NFIP, however, is not whether hurricane risks are higher in the near term (because of cyclical factors that may be well-represented in the decades of historical data underlying FEMA's models).

What matters for the NFIP is whether climate change has increased hurricane risks in the long term, averaged over cyclical ups and downs. There is less agreement on that question.[38] On the one hand, data for the North Atlantic from 1950 through 2005 show a very strong correlation between the Power Dissipation Index and sea surface temperatures, and several studies have concluded that those temperatures have risen partly because of greater concentrations of greenhouse gases in the atmosphere. Moreover, statistical analysis suggests significant increases in the number of North Atlantic hurricanes if the initial year of the period analyzed is between 1881 and 1921, suggesting that a long-term trend is at work. On the other hand, hurricane counts were also high between 1851 and 1871, and no strong trends are found in analyses that include those years. Nor is there evidence of a trend in the number of hurricanes that make landfall in the United States.[39]

One possible explanation for that lack of clear evidence is that climate change has had only a negligible impact, if any, on hurricane risks in the United States. Temperatures in the North Atlantic—where hurricanes striking the U.S. Atlantic and Gulf coasts form—have risen by about 0.67 degrees Celsius (1.2 degrees Fahrenheit) since 1906, researchers estimate. They attribute just over 40 percent of that increase to global warming.[40] A review of studies suggests that for each rise of 1 degree Celsius in average sea surface temperature, wind speeds of the strongest hurricanes increase between 1 percent and 8 percent, all other things being equal.[41] The destructive power of a hurricane grows with the cube of the wind speed, so the increase of 40 percent of 0.67 degrees Celsius attributable to global warming translates to an increase of about 1 percent to 7 percent in the destructive power of storms. That effect is relatively small, and it could be less important than a second, offsetting impact of climate change on hurricane formation: an increase in vertical wind shear (changes in wind speed or direction as altitude changes). Vertical wind shear tends to prevent weak disturbances from intensifying into hurricanes and may limit the intensity of storms that have already formed.

Another possible explanation for the current inconclusive evidence is that climate change is already having an important impact on hurricane risks in the United States but more data must accumulate before that effect can be confirmed statistically.[42] In that view, the influence of climate change is being obscured for now by the high variability of hurricane events from year to year and by large-scale quasi-periodic oscillations in weather patterns that influence hurricane activity in the United States, such as El Niño, West African monsoons, and perhaps the Atlantic Multidecadal Oscillation.[43]

If the various effects of climate change on hurricane risks do not neutralize each other, their net impact could become more pronounced over time as carbon dioxide and other greenhouse gases continue to accumulate in the atmosphere. Most global climate models indicate that sea surface temperatures and maximum wind speeds will continue to rise in the future.[44] Many also anticipate increases in accompanying rainfall in some areas. William Nordhaus has estimated that if the atmospheric concentration of carbon dioxide (or its equivalent in other gases) doubled, the mean effect would be to roughly double expected U.S. losses from hurricanes—an increase of about $8 billion in 2005 dollars.[45] Moreover, applying a recently developed analytic technique to results from climate models indicates that global warming could reduce the frequency of hurricanes worldwide but increase their frequency off the southeastern coast of the United States.[46] That analysis also concluded that the intensity of hurricanes could increase in some locations, including the North Atlantic. However, even if climate change had no significant effect on the number or intensity of hurricanes reaching the United States, it would still increase flood risks in coastal areas through its effect on sea level.

Private Insurers' Response to Climate Change

Some support for the view that climate change may have a significant impact on flood risks, now or in the future, comes from the concern seen among private insurers who provide coverage for flood or wind damage in areas subject to hurricanes (see Box 2 on page 8). For example, a coalition of insurance companies, environmental groups, and other organizations recently issued a report on risks to U.S. coastal communities that said, "Changing climatic conditions pose an unprecedented threat to U.S. coastlines. . . . Current flood, shoreline and inundation maps, used for land use and infrastructure planning and mortgage due diligence, do not accurately reflect current risks, let alone future risks."[47] A 2006 report from insurer Lloyd's said, "Insurers should plan for increased flood risk in areas expected to suffer greater rainfall."[48]

What private insurers are already doing to respond to the effects of climate change is largely unknown. For a 2007 report, the Government Accountability Office contacted representatives of 11 large property casualty insurers; all 11 said their firms were using models that predicted greater frequency and intensity of hurricanes in the near term, but they did not attribute those effects to climate change.[49] Two industry sources told CBO that some private flood insurers have tightened their eligibility criteria for various combinations of reasons, which may include climate change. However, a representative of a firm that produces flood maps in Europe told CBO that the company does not yet have enough data to justify specific changes to its models of flood frequency and severity.

Options to Address Concerns About the NFIP's Full-Risk Rates

Various lawmakers and interest groups have proposed changes to the policies governing the NFIP. Some proposals would address the sources of actuarial weakness that CBO has identified; others would not.

With regard to the problem that FEMA's older maps may understate the damages expected from a given weather event, possible responses include providing more funds to speed the agency's Map Modernization program or to shorten the interval between map reviews once the modernization effort is completed. Updating flood maps more often would allow the rates charged on new policies to reflect more-current floodplain boundaries and base flood elevations. (Under FEMA's present grandfathering policy, new information would not alter the rates charged on existing policies.) Another option would be to let FEMA base its rates not on currently observed conditions—such as land use, location of shoreline, and so forth—but on the average conditions expected between one map review and the next.

FEMA is conducting an analysis to address whether its estimates of flood frequencies are too low because they do not account for the effects of climate change. The agency hopes to finish that analysis early next year. Depending on the conclusions of the study, it may seem worthwhile to establish a regular periodic review so that FEMA can stay abreast of the latest research findings on climate change.

To address concerns about FEMA's ability to respond quickly to new data showing higher flood risks, the 10 percent cap on annual premium increases could be lifted. At present, the cap is rarely a binding constraint on the rates that FEMA's actuaries would choose to set to cover current flood risks. As explained above, current rates in the V zones are actually higher because of the cap than they would be otherwise. However, lifting the cap could help improve the NFIP's financial position in the future if FEMA's analysis of the effects of climate change indicates a need for large increases in some rates or if erosion leads to rapid growth in the risks to coastal properties.

Another concern is the grandfathering of properties that otherwise would have moved from a regular 100-year floodplain (A zone) to an area subject to velocity wave action (V zone) or that would have been reclassified at a lower elevation relative to the base flood. Lawmakers could address that issue by prohibiting such grandfathering. Alternatively, they could require FEMA to track the premium income lost through that practice and offset it by raising rates elsewhere, thus creating a cross-subsidy.

Grandfathering properties that had been in Zone X could also be prohibited, but doing so would have little or no impact on the NFIP's actuarial balance, because those properties are already cross-subsidized. The same is true of policy actions to reduce the residual risks to properties protected by levees (see Appendix B) or to adjust the floodplain boundaries on FEMA's maps to correct for small-sample bias in communities with relatively short data records (see Appendix A).

Some other common proposals also would not address the sources of actuarial weakness in the NFIP's rates. An example is proposals to expand the program's coverage— for instance, by raising coverage limits or by adding coverage for wind damage. Proposals envision offering the additional coverage at actuarially neutral rates, but meeting that goal would be difficult, as it is for the program's current coverage. Raising coverage limits, for instance, would disproportionately affect properties in coastal states; therefore, the actuarial balance of the premiums would be particularly sensitive to any weakness in FEMA's modeling of hurricanes.[50] Even if the premiums for the new coverage were actuarially fair, the program's overall actuarial balance would remain the same in dollar terms.[51]

A full analysis of policy options for the NFIP is beyond the scope of this report, and it should be noted that actuarial soundness is not the only goal policymakers have set for the program. Even from the standpoint of federal spending, the effect of the program depends not only on its own financial position but also on its interaction with federal disaster assistance: The more properties that are adequately insured by the NFIP, the fewer applications will be filed for disaster assistance from FEMA or low-cost loans from the Small Business Administration after the President declares a flooding disaster. Quantifying the extent of that effect is one of several research goals that could help inform future policy choices about the NFIP (see Box 3).

Beyond its federal fiscal impact, the NFIP is important for its effects on environmental protection, public safety, and economic efficiency. Expanding floodplain boundaries where needed to correct for small-sample bias would be desirable on some of those grounds, because applying the program's minimum-elevation requirements for new construction as well as other land-use standards would decrease future flood damage. Reducing cross-subsidies could also be beneficial if policyholders reacted to rates that reflected their properties' true risks by taking cost-effective actions to mitigate those risks. In both cases, however, the benefits could be reduced if participation in the program fell.

NFIP Properties That Have Experienced Multiple Floods

Another concern that some observers of the NFIP have raised is the impact of insured properties that have flooded more than once. Such properties are responsible for only a small share of claims paid by the NFIP, but they impose a disproportionate burden on the program's finances.

The Number and Effects of Repetitive-Loss Properties

About 71,000 properties insured by the NFIP as of January 31, 2009, meet FEMA's definition of repetitive-loss properties—that is, they have been the subject of two or more flood-claims payments greater than $1,000 each in any 10-year period.[52] Nearly 32,000, or 45 percent, of those RLPs have prompted three or more such payments while insured by the program, and a few thousand have had losses leading to least 10 payments (see Figure 2). Properties with the largest numbers of claims payments typically experience minor damage, such as ruined carpeting, each time. Thus, of the almost 23,000 properties that have prompted at least four payments, only about 10,000 meet FEMA's criteria for a severe repetitive-loss property (SRLP)—that is, they have been the subject of at least four payments of more than $5,000 each, or two or more payments that cumulatively exceed the property's value.[53]

Box 3. Questions for Further Research on the Demand for Flood Insurance

If the premium rates charged on some or all groups of policies in the National Flood Insurance Program (NFIP) are too low to cover their expected costs, raising those rates would improve the program's actuarial soundness and lessen the need for future borrowing from the federal Treasury. Raising rates, however, would have other effects that can be described in qualitative terms but are not well understood quantitatively. Improved answers to the following questions could help the Congress make more informed policy decisions about the NFIP:

- To what extent would raising premiums reduce the number of policies in force or the amount of coverage per policy?
- To what extent would raising premiums increase expected losses per policy by disproportionately reducing the demand for coverage of lower-risk properties?
- To what extent would raising premiums lead to greater mitigation—for example, by encouraging policyholders to elevate their properties or move to less risky locations?
- To what extent would raising premiums increase federal costs for disaster assistance?

There are few empirical studies of the demand for flood insurance, and none of them provide definitive answers. The best available evidence suggests that raising premium rates by 10 percent lowers the number of policies by 1 percent and the amount of coverage per policy by 9 percent. However, those estimates were derived from an analysis of aggregate state-level data from 1984 through 1993.[1] Thus, they do not reflect the effects of the changes made by the National Flood Insurance Reform Act of 1994 to the requirement for flood coverage of certain structures in high-risk areas. (Under that law, federally regulated mortgage lenders and government-sponsored enterprises such as Fannie Mae and Freddie Mac must put NFIP premium payments in escrow for any loan that has an escrow account for other purposes.) Moreover, the degree to which premium rates affect the demand for flood insurance may be different at the current rates than it was during the 1984–1993 period.

Another study used 1997 NFIP data on 62 coastal communities in 16 counties to analyze the percentage of properties covered in each community. It did not find statistically significant evidence that the average price of NFIP insurance has any effect on the extent of coverage, perhaps because most covered properties were subject to the mandatory-purchase requirement.[2] A third study was commissioned by the Federal Emergency Management Agency to analyze the potential effects of eliminating the explicit subsidy for structures built before their community's flood insurance rate map was created. However, the methodology of that study focused on surveying structures in a nationwide sample of communities with high-risk flood zones; it did not include any new analysis of how the demand for flood insurance is affected by price, instead relying on a single 1983 report by the Government Accountability Office.[3]

[1] Mark J. Browne and Robert E. Hoyt, "The Demand for Flood Insurance: Empirical Evidence," *Journal of Risk and Uncertainty*, vol. 20, no. 3 (2000), pp. 291–306.

2 Craig Landry and Warren Kreisel, "Modeling the Decision to Buy Flood Insurance: An Empirical Analysis for Coastal Areas" (paper presented at the American Agricultural Economics Association annual meeting, Tampa, July 30– August 2, 2000), available at http://purl.umn.edu/21880.

[3] PricewaterhouseCoopers, *Study of the Economic Effects of Charging Actuarially Based Premium Rates for Pre-FIRM Structures* (report prepared for the Federal Emergency Management Agency, May 14, 1999), available at www.fema.gov/library/viewRecord.do?id=2555. The 1983 report used in that study was General Accounting Office (now Government Accountability Office), *The Effect of Premium Increases on Achieving the National Flood Insurance Program's Objectives,* GAO/RCED-83-107 (February 28, 1983). Also see Congressional Budget Office, letter to the Honorable Rick A. Lazio regarding CBO's review of a study of the economic effects of charging actuarially based premium rates for federal flood insurance, September 22, 2000.

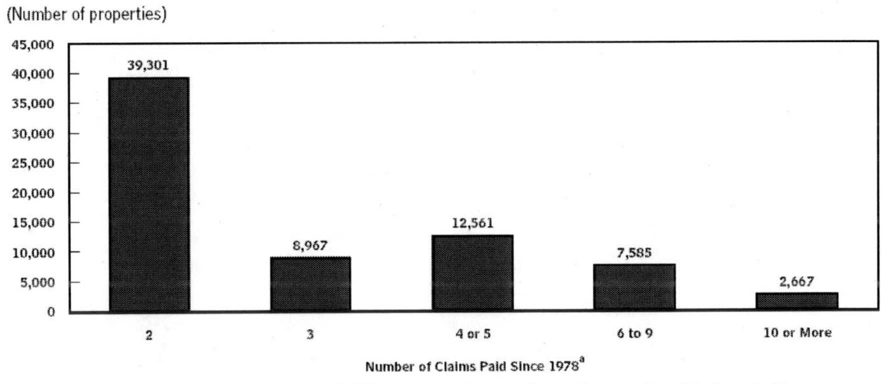

Source: Congressional Budget Office based on data from the Federal Emergency Management Agency as of January 31, 2009.

[a] Includes only payments that exceed $1,000.

Figure 2. Currently Insured Repetitive-Loss Properties, by Number of Claims Paid.

The 71,000 currently insured RLPs are covered by a total of about 110,000 policies, or 2 percent of the 5.6 million NFIP policies nationwide (see Table 2).[54] RLPs account for 3 percent of current NFIP premiums, indicating that the rates on individual policies tend to be 50 percent higher than the national average. However, currently insured repetitive-loss properties account for 12 percent of the total number of claims paid by the NFIP since 1978 and 16 percent of total claims payments (in nominal dollars). SRLPs have an even more disproportionate impact on the program's actuarial balance: They represent just 0.3 percent of current policies and 0.5 percent of current premiums, but 3 percent of claims and 5 percent of payments since 1978.

Table 2. Statistics on Insured Properties with Multiple Flood Claims

	Currently Insured Properties	
	RLPs	SRLPs
Number of Properties	71,081	10,054
Number of Policies	109,842	17,814
Percentage of Current Policies	2.0	0.3
Percentage of Current Premiums[a]	3.1	0.5
Percentage of Cumulative Claims[b]	12.5	3.3
Percentage of Cumulative Claims Payments[b]	15.7	5.1
Percentage of Policies in High-Risk Zones That Are Subsidized[c]	56.2	62.3
Percentage of Policies in V Zones[d]	8.3	9.7

Source: Congressional Budget Office based on data from the Federal Emergency Management Agency as of January 31, 2009, and on Thomas L. Hayes and Dan R. Spafford, *Actuarial Rate Review: In Support of the May 1, 2008, Rate and Rule Changes* (Federal Emergency Management Agency, National Flood Insurance Program, August 2008).

Notes: Repetitive-loss properties (RLPs) are properties that have been the subject of at least two flood-claims payments of more than $1,000 each in any 10-year period. (The numbers shown here for RLPs include severe repetitive-loss properties.)

Severe repetitive-loss properties (SRLPs) are RLPs that have prompted four or more flood-claims payments greater than $5,000 each, or two or more payments that together exceed the property's value. (The numbers shown here for SRLPs include 1,195 nonresidential properties, although such properties are not defined as SRLPs for purposes of mitigation assistance under the pilot program created by the Flood Insurance Reform Act of 2004.)

[a] Current premiums are shown because data on cumulative premiums are not available.

[b] Cumulative data for January 1, 1978, through January 31, 2009.

c The analogous figure for the flood insurance program as a whole is 35.4 percent.

[d] V zones are coastal areas subject to waves of three feet or more during a 100-year flood. The analogous figure for the flood insurance program as a whole is about 2 percent.

Looking only at the losses to date on currently insured RLPs may understate the significance of the repetitive-loss problem. Over time, the current set of properties will experience some additional losses, and additional properties will become RLPs. Another way to measure the impact of repetitive-loss properties is to include former RLPs—those that are no longer insured or are no longer considered active RLPs because they have undergone mitigation to reduce their risks. Including those properties raises RLPs' share of total claims since 1978 from 12 percent to 22 percent and their share of total payments from 16 percent to 24 percent.[55]

In high-risk zones, RLPs are much more likely than all NFIP properties (56 percent versus 35 percent) to be insured at subsidized rates. That situation reflects the fact that properties eligible for subsidies are older ones that were built before communities joined the NFIP and had to meet the program's standards for building codes and floodplain management. Also, policies on RLPs are about four times as likely as all policies to cover properties in V zones. For SRLPs, the contrasts are sharper still.

The apparent imbalance for repetitive-loss properties between current premiums and cumulative claims, however, cannot be taken at face value as indicating the effect of an individual RLP on the actuarial soundness of the NFIP. A property that has experienced two floods in a 10-year period may indeed be at a higher risk of future flooding, or it may simply have been subject to a run of bad luck despite having an average or below-average level of risk. One cannot assume that the owners of such properties would pay higher actuarially sound premiums any more than one can assume that the owners of properties that have experienced no losses over 10 years face no flood risks. The statistics for SRLPs are more meaningful: A series of four claims payments of more than $5,000 each (or two to three payments that together exceed the property's value) is unlikely to reflect merely a run of bad luck.[56]

FEMA's Treatment of Repetitive-Loss Properties

In dealing with RLPs, FEMA's main focus is on efforts to reduce their risk of future flooding, not on their premiums. NFIP premiums are not based on the loss experience of individual properties, and most RLPs are not targeted for special rates as a group. Although the average premium paid on policies that insure RLPs is higher than in the NFIP as a whole, that is mainly because those policies are more concentrated in the high-risk coastal and subsidized categories.[57] RLPs' loss experience affects their rates in one case, however: Repetitive-loss properties located in Zone X are ineligible for the discounted preferred-risk policy premiums that apply to most properties outside 100-year floodplains. Thus, RLPs in Zone X are insured at premium rates well above average for their area (see Table 3).

Table 3. Average Annual Premiums, by Risk Zone, for All Properties and for Multiple-Loss Properties

(Dollars)			
	All Properties	RLPs	SRLPs
	Full-Risk Premium Rates		
A Zones (High risk: annual flooding probability of 1 percent or more)	430	467	484
V Zones (High risk coastal, subject to 3-foot waves in a 100-year flood)	2,270	1,389	950
Zone X (Low to moderate risk)	394	640	504
	Subsidized Premium Rates		
A Zones (High risk: annual flooding probability of 1 percent or more)	980[a]	1,140	1,024
V Zones (High risk coastal, subject to 3-foot waves in a 100-year flood)	1,500[a]	1,506	1,253

Source: Congressional Budget Office based on data from the Federal Emergency Management Agency.

Note: Repetitive-loss properties (RLPs) are properties that have been the subject of at least two flood-claims payments of more than $1,000 each in any 10-year period. Severe repetitive-loss properties (SRLPs) are RLPs that have prompted four or more flood-claims payments greater than $5,000 each, or two or more payments that together exceed the property's value.

[a] Approximation based on Thomas L. Hayes and Dan R. Spafford, *Actuarial Rate Review: In Support of the May 1, 2008, Rate and Rule Changes* (Federal Emergency Management Agency, National Flood Insurance Program, August 2008). Subsidized rates are available only in high-risk zones, not in Zone X.

Conversely, among properties in V zones that are insured at full-risk rates, RLPs and SRLPs pay premiums that are much lower than average. That situation may result from the greater prevalence of condominium policies among multiple-loss properties; such policies can be less costly because the risk is shared among a larger number of owners.[58]

An RLP's risk of future flooding may be mitigated in several ways: by elevating the structure; by acquiring the structure, relocating or demolishing it, and converting the land to open-space uses; and, in some cases, by dry flood-proofing (sealing the exterior walls to prevent water from entering) or building localized physical flood- control measures (levees or floodwalls). FEMA provides grants to help state and local governments reduce flood risks through several programs that have varying eligibilities and requirements, including the Hazard Mitigation Grant Program, the Flood Mitigation Assistance Program, and the Pre-Disaster Mitigation Program (see Box 1 on page 4).

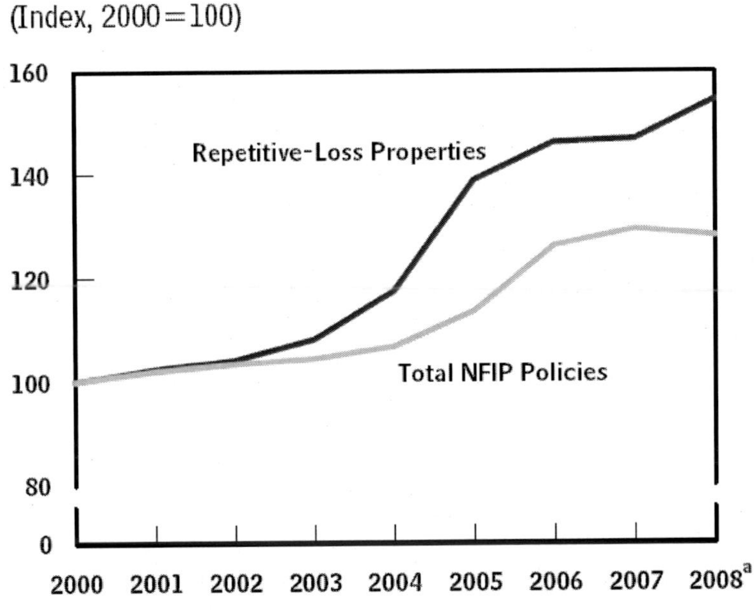

Source: Congressional Budget Office based on data from the Federal Emergency Management Agency.

Notes: NFIP = National Flood Insurance Program.

Repetitive-loss properties are properties that have been the subject of two or more flood-claims payments greater than $1,000 each in any 10-year period.

[a] Data for 2008 include January 2009.

Figure 3. Growth in the Number of Repetitive-Loss Properties and Total NFIP Policies Since 2000.

Legislation enacted in 2004 established two additional programs focusing on RLPs: the Repetitive Flood Claims (RFC) Grant Program and the Severe Repetitive Loss Program.[59]

- RFC grants are targeted to state or local governments that lack the financial or managerial capacity to take part in the Flood Mitigation Assistance Program. Those grants do not require any local sharing of costs. Through December 2007, 66 repetitive-loss properties had been mitigated via RFC grants, and another 13 projects had been approved but not completed.
- The Severe Repetitive Loss Program includes both a carrot and a stick. An SRLP insured at subsidized rates can receive mitigation assistance under the program, but if the property owner rejects the mitigation offer, the NFIP premium rises by 50 percent at the next policy renewal and by another 50 percent after each subsequent flood claim, until the full-risk rate is reached. FEMA first issued guidance for the new program in January 2008. As of September 30, 2008, grants had been awarded for mitigation of 168 properties, 132 of them in Texas. Mitigating all of the residential SRLPs in the nation by acquiring them—the typical method for such properties—would cost $1.8 billion (including a nonfederal cost share of 10 percent to 25 percent), according to the Department of Homeland Security.[60] That estimate was based on 8,040 residential SRLPs as of December 2007. The comparable figure for the end of January 2009 is $2.0 billion—nearly 10 times the total appropriations for the Severe Repetitive Loss Program through fiscal year 2009.

Despite mitigation efforts by FEMA and state and local governments, the number of repetitive-loss properties insured by the NFIP has grown by more than 50 percent in this decade: from 45,783 at the beginning of 2000 to 71,081 as of January 31, 2009. Annual growth rates for such properties have been roughly consistent with growth in the total number of NFIP policies, except in 2004 and 2005 (see Figure 3). The more rapid growth of RLPs in those years is largely associated with the effects of the hurricanes of 2004 and 2005. Annual growth is measured as the difference between the number of insured properties that newly meet the definition of an RLP and the number of properties that cease to be insured (because they have been destroyed or abandoned or the policyholder has dropped coverage) or that have undergone mitigation and are therefore not considered active RLPs.[61] The number of properties leaving the active list fell sharply in 2005 and 2006—perhaps because public efforts were focused on short-term cleanup and restoration after the hurricanes rather than on mitigation—but rose again in 2007 and 2008 (see Figure 4).

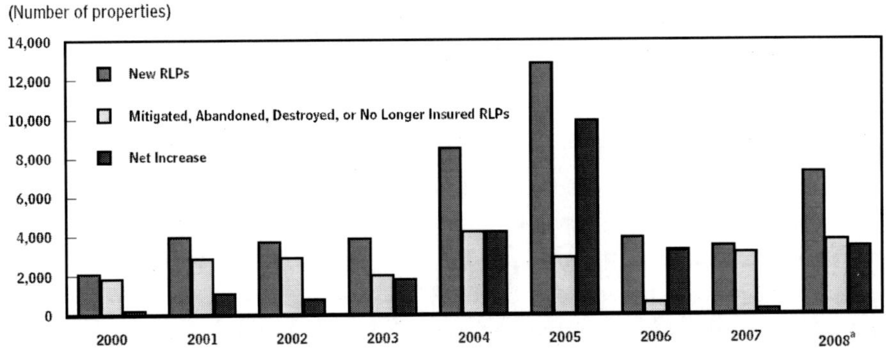

Source: Congressional Budget Office based on data from the Federal Emergency Management Agency.

Note: Repetitive-loss properties (RLPs) are properties that have been the subject of two or more flood-claims payments greater than $1,000 each in any 10-year period.

[a] Data for 2008 include January 2009.

Figure 4. Annual Changes in the Number of Insured Repetitive-Loss Properties.

APPENDIX A: FURTHER DETAILS ABOUT THE NFIP'S FULL-RISK RATES

This appendix presents supplementary information about the zone system and rate-setting methods used in the National Flood Insurance Program (NFIP).

Classification of High-Risk Flood Zones

In drawing up its flood insurance rate maps, the Federal Emergency Management Agency (FEMA) delineates three major types of flood zones. Most coastal or inland areas in 100-year floodplains—that is, places where the annual probability of flooding is at least 1 percent—are labeled A zones. However, coastal areas where a 100-year flood would include three-foot waves on top of the storm surge itself are labeled V zones (for "velocity wave action"). Areas outside 100-year floodplains are designated Zone X.[1]

FEMA's rate structure can be more detailed in areas for which it has more information, so the agency breaks down its zone categories further according to the level of analysis it has performed. In the V zones, more than 99 percent of policies with full-risk premiums cover properties in areas for which detailed analyses are available. Under an older rating system, those areas carried various numbers from V1 through V30, but now they are all labeled Zone VE. The relatively few V-zone areas that have been mapped more approximately are labeled Zone V—sometimes referred to as "unnumbered Zone V" to avoid confusion between that label and the larger V-zone category (see Table A-1).

Areas in the A zones are similarly classified according to the level of information available, and FEMA also distinguishes areas that are subject only to shallow flooding. In particular, Zone AE (formerly Zones A1 through A30) refers to areas studied in more detail, and ("unnumbered") Zone A is the label used for areas subject to normal (not merely shallow) flooding that have been mapped more approximately. In addition, Zone AH designates areas where flooding takes the form of shallow ponding— for example, on a plateau where flood waters can reach only one foot before they start spilling off the plateau. Zone AO covers other areas that are subject to shallow flooding, usually in the form of sheet flow (shallow flow over a wide area) on sloping terrain. Among all full-risk policies in those four A zones, roughly 75 percent cover properties in Zone AE, 5 percent in Zone A, and 20 percent in Zone AH or AO.[2]

As discussed below, FEMA uses models to set full-risk rates for policies that cover properties in Zones VE and AE, where it has more-detailed information about flood risks. The agency has less-detailed information about risks in Zones V, A, AH, and AO, so its rate setting for those areas relies more on professional judgment. For the relatively few properties in unnumbered Zone V, FEMA's actuaries determine premium rates on a case-by-case basis.[3] For unnumbered Zone A, three sets of rates are used. One set applies to buildings whose position relative to the base flood elevation (BFE)—that is, the height of water reached during the local 100-year, or base, flood— has been certified.[4] A second set of rates applies to structures for which the elevation relative to the adjacent grade (ground level) is specified. A third set applies to buildings for which no elevation information is available. FEMA's actuaries calculate those three sets of rates using Zone AE rates as a benchmark. For example, the BFE-rated schedule is somewhat higher than the comparable schedule in Zone AE, and the adjacent-grade-rated schedule assumes that the ground level is two to three feet below BFE. For areas subject only to shallow flooding (Zones AH and AO), actuaries use a different set of rates; for any given type of structure, those rates depend only on whether the structure is certified as lying above the community's BFE.

Table A-1. Distribution of Full-Risk Flood Insurance Policies by Zone

Major Zone	Subzone		Description	Share of Full-Risk Policies (Percent)
V			Coastal areas subject to waves of three feet or more during a 100-year flood	
	VE		V-zone areas mapped using detailed methods	1.3
	V (or "unnumbered V")		V-zone areas mapped using approximate methods	___ *
		Total V		1.3
A			All other coastal and inland areas in	
			100-year floodplains	
	AE		A-zone areas mapped using detailed methods	37.2
	A (or "unnumbered A")		A-zone areas subject to more than shallow flooding	
			and mapped using approximate methods	2.2
	AH, AO		Areas subject to shallow ponding (AH) or	
			other shallow flooding (AO)	10.6
		Total A		50.0
X	(No subzones)		Areas outside 100-year floodplains	48.8

Source: Congressional Budget Office based on estimates for 2008 to 2009 in Thomas L. Hayes and D. Andrew Neal, *Actuarial Rate Review: In Support of the Recommended May 1, 2009, Rate and Rule Changes* (Federal Emergency Management Agency, National Flood Insurance Program, August 2009), p. 35.

Note: * = less than 0.05 percent.

The Role of Representative Topographic Profiles

Ideally, the full-risk rates charged in Zones AE and VE, where FEMA has the most detailed information, would reflect not only each structure's type and its location relative to the base flood elevation but also its local topography. Depending on local conditions, structures at the same elevation relative to

BFE—and hence exposed to roughly the same depth of water (if any) in a 100-year flood—may face different risks from floods of other sizes. For instance, a house that is one foot above BFE may be safe from all floods smaller than a 300-year event if it is located in a broad, shallow floodplain, where the water does not rise much higher in a rare large flood than in a more common small one. Conversely, a house located one foot above BFE in a narrow, steep valley, where the difference in water heights between rare and common floods is greater, may suffer damage in all floods larger than a 130-year event.

Basing flood insurance rates on each structure's local topography is not currently feasible, however, because of the information requirements and administrative complexity involved. Instead, FEMA uses estimates of average topographic conditions to set uniform nationwide rates for Zone AE and for Zone VE. The agency derives those rates by averaging six representative topographic profiles, summarized by "probability of elevation" (PELV) curves. Each PELV curve shows the relationship for a particular topography between flood recurrence intervals (the reciprocals of the annual probabilities) and flood heights, which are expressed relative to BFE (for example, BFE +2 feet). All six curves intersect at the common point (100, BFE)—because the base flood is the 100-year flood by definition—but they differ in how steeply or gradually they slope away from that point. In particular, each curve shows a different height below BFE for a 10-year flood. Indeed, FEMA derived the curves by analyzing the flood exceedence probabilities for a sample of communities whose 10-year floods share the same relative height.[5]

To determine how to weight the six curves in calculating the national averages, FEMA uses state-by-state data on the distribution of 10-year flood heights relative to BFE for some of its insured properties. (Initially, FEMA's maps for the A and V zones showed subzones that indicated the relative height of the 10-year flood at each location, but current maps do not always include that information.) For instance, if 19 percent of the properties for which FEMA has data in a given state are located in places where the 10-year flood has a height of about BFE-3 feet, the agency assumes that the locations of 19 percent of all insured properties in that state fit the topographic profile represented by the PELV curve that includes the point (10, BFE -3).

The PELV curves therefore allow FEMA to reduce its information requirements, generalizing from data on the relative height of 10-year floods to the heights of floods of all recurrence intervals—or conversely, to the exceedence probabilities for floods of all relative heights. Combining the probabilities using each state's share of insured properties and the respective weights for each PELV curve in the state yields a set of average probabilities for the nation as a whole. FEMA can then estimate expected percentage losses for each zone (AE or VE), structure type, and structure elevation relative to BFE by combining those average probabilities of flood heights with a set of damage functions—one for each zone and structure type—that relate the depth of flooding in a structure to an estimated percentage loss. The damage functions are zone-specific because the function for Zone VE takes account of wave action and effects on piers and pilings below the lowest floor.[6]

FEMA bases the damage functions on a combination of models from the Army Corps of Engineers and its average claims experience.[7] In calculating its average experience, the agency applies a weight of 1 percent to the data from 2005 and a weight of 99 percent to the average of all other years, on the grounds that the long-duration flooding of 2005 was highly unusual and should not excessively influence future expectations. Because FEMA now has about 30 years of claims data, going back to 1978, those weights have the effect of reducing the impact of the 2005 experience from about 3 percent (1/30) to 1 percent—in effect, treating 2005 as a 1-in-100-year event. The agency has commissioned a study to provide recommendations on the appropriate treatment of the 2005 experience.

FEMA's Correction for Short Historical Records in Zone AE

In some cases, the method outlined above for calculating expected percentage losses underestimates the actual risks: A small-sample bias exists whenever the length of available data is short relative to the return period of the risks being studied. For example, FEMA estimates that for a location with 25 years of data, a 100-year flood estimated without correcting for sample bias can be expected to have a true return interval of 63 years—or, in other words, to have an annual exceedence probability of 1.59 percent (1/63) instead of the intended 1 percent.

The many small samples that exist nationwide include representative numbers of small and large floods. But the errors in estimated flood risks—with some estimates being too high and some too low—do not cancel each other out. FEMA uses flood records and a statistical model of flood flow in a local area to calculate an exceedence probability. Graphically, that probability is a decreasing and concave function of flood flow for rare floods. In FEMA's modeling, areas that have below-average flood experience exert disproportionate influence on the estimated national average probability compared with areas that have above-average flood experience. Intuitively, for a true exceedence probability of 1 percent, for example, estimation errors from small samples resulting in low estimates must be between zero and 1 percent, while errors resulting in high estimates may be much greater than 1 percent—and the average of the high and low estimates will be greater than 1 percent. A second but smaller source of bias from short historical records is downward bias in estimates of sample standard deviations.[8]

Since the late 1980s, FEMA has compensated for small-sample bias in its estimates for Zone AE by conducting the above analysis again using a second set of elevation probability curves that incorporate a correction for small samples.[9] Those curves, called PELV(500) curves, assume that the short data records cover 25 years.[10] FEMA combines the results obtained using the two sets of PELV curves; in doing so, it uses weights of 20 percent for the PELV(500) results and 80 percent for the regular PELV results, based on its original estimate that 20 percent of the NFIP's policies were in communities with short records.

If those weights produced the right amount of correction for small-sample bias when FEMA first implemented the approach in the late 1980s, they may be leading to overcompensation now and thus contributing to an actuarial surplus on policies in Zone AE (and by extension in unnumbered Zone A). One reason is that the percentage of the NFIP's exposure that is in communities with relatively short records has declined in the past two decades as the program's policies have become increasingly concentrated in coastal areas, where flood records tend to be longer. Another reason is that over the same period, communities with shorter records have accumulated additional data, some of which have been incorporated into FEMA's models when flood maps have been revised.

FEMA's approach to the problem of small-sample bias has some important limitations. First, it creates a cross-subsidy because it raises the rates charged on all Zone AE policies nationwide, not just on those in communities with short flood records. Second, small-sample bias is a factor that shrinks the areas mapped as 100-year flood-plains and reduces the estimated depth of flooding in communities with short historical records. All else being equal, in a community that has 25 years of data, what appears to be a 1-in-100 chance of flooding on the basis of those data can be expected, on average, to be a 1-in-63 chance. For such a community, therefore, the mapped 100-year floodplain could actually be a 63-year flood-plain, making the annual risk of flooding in Zone X just outside the floodplain boundary not 1 percent but 1.59 percent. Some other factors in FEMA's mapping process, however, also have not been updated for many years.[11] Those factors could exacerbate or offset the effect of the small-sample bias.

Undersized floodplains and underestimated flood depths in communities with short historical records would negatively affect the NFIP's floodplain management goals by reducing the effectiveness of the program's elevation requirements for new construction in 100-year flood-plains. Specifically, when floodplains are mapped too small, the requirements do not apply in some areas where they would otherwise; and where they do apply, the requirements are lower than intended.

The mapped errors, however, probably have little or no adverse impact on the actuarial soundness of the NFIP. One reason is that the "additional" floods that affect properties in the mapped A zones because the elevation requirements are too low should be offset (or more than offset) by the higher premium rates that result from the PELV(500) adjustment. As for the properties that lie outside a mapped floodplain even though their true flood risk exceeds 1 percent, they are insured at rates that generally do not cover their full expected costs, but the difference is funded largely or entirely through cross-subsidies from other Zone X policyholders, not taxpayers. As discussed in the main text, the starting point for FEMA's estimate of expected losses for structures insured at Zone X standard rates is the observed losses on those policies (which FEMA then multiplies by the ratio of estimated to observed losses for Zone AE). Therefore, if higher flood risks for properties outside an undersized floodplain lead to greater losses on those properties, they drive up the rates paid by all Zone X standard policyholders. The connection between undersized floodplains and higher rates is less automatic for preferred-risk policies in Zone X, because FEMA's actuaries do not set those rates according to a simple formula. Nevertheless, a connection between greater risks and higher rates is likely to occur over time as the actuaries apply their professional judgment to the available data. (For more details about cross-subsidies in the NFIP, see Appendix B.)

Converting Expected Percentage Losses into Premium Rates

For insured properties in Zone AE, after estimating expected percentage losses using the PELV and PELV(500) curves discussed above, FEMA converts those loss estimates into premium rates by using several adjustment factors. In keeping with standard insurance practices, some of the adjustments take into account administrative expenses, optional deductibles, loss-adjustment expenses, and underinsurance (that is, coverage for less than the full value of the insured property). In addition, FEMA includes a contingency loading of 10 percent to provide a safety margin for uncertainty, and it offsets the discounts given under the Community Rating System (CRS) by scaling the base (undiscounted) rates upward so as to keep the CRS program revenue neutral. Those various adjustment factors, other than the contingency loading, are reviewed by FEMA each year and revised as necessary.[12]

FEMA also adjusts the rates as necessary to keep them between certain lower and upper bounds. Specifically, it ensures that rates do not fall below minimum levels chosen to cover the agency's fixed expenses per policy and that average rates for each type of structure do not rise by more than 10 percent a year.

For properties in Zone VE, FEMA uses the same process to convert expected percentage losses into premium rates, with two exceptions. First, it currently applies a contingency loading of 20 percent, not 10 percent. Second, since 2001, FEMA has generally added one more step to scale the rates upward to the extent allowed by the 10 percent cap on annual increases.[13] As discussed in the main text, that extra step is the agency's response to the costs associated with coastal erosion and to the cap itself. Erosion-related costs are expected to rise so rapidly in the future as shorelines continue to move inward, toward areas of denser development, that the 10 percent cap would not allow FEMA's premium income from Zone VE to keep pace. Raising current rates above the levels thought to be sufficient to cover current costs prefunds some of the future risk. It also allows future rate increases to be larger in dollar terms than they would be otherwise because the 10 percent cap will apply to a higher base.

APPENDIX B: RESIDUAL RISK BEHIND LEVEES AND OTHER CROSS-SUBSIDIZED RISKS

Across-subsidy in insurance coverage exists whenever the rates charged on some policies exceed the expected cost of the claims and expenses on those policies (including the return on capital) while the rates charged on other policies fall below their expected cost. In practice, no insurance program can be completely free of cross-subsidies because insurers cannot collect information on all of the myriad factors that influence the level of risk. When rates are based on past claims, every omitted factor that increases risk creates a cross-subsidy from the policyholders who do not have that factor to those who do (and conversely for omitted factors that reduce risk). In the case of the National Flood Insurance Program (NFIP), unpriced risk factors create cross-subsidies to the extent that they affect flood risks to properties outside 100-year floodplains (that is, in Zone X) or affect the damage caused by a given depth of flooding in any zone—both of which the Federal Emergency Management Agency (FEMA) estimates on the basis of the program's experience.[1]

Cross-subsidies do not necessarily undercut the actuarial soundness of the NFIP, but they can decrease economic efficiency. Policyholders who pay rates that do not fully reflect their risks have less financial incentive to take cost-effective actions to reduce those risks. That effect applies whether or not the reduced rates are cross-subsidized. The other adverse impact of cross-subsidies is that potential policyholders who face premiums that are higher than they would be otherwise may choose to buy less coverage or forgo it entirely. Conceivably, the incentives to reduce or avoid those higher rates could also encourage some property owners to take excessive measures (ones whose costs exceed the expected benefits) to reduce their risks if doing so would lower their premiums.

An important cross-subsidy in the NFIP involves policies that cover properties in areas protected by levees, dams, walls, or other flood-control structures. (For simplicity, such structures are referred to collectively as levees hereafter.) This appendix explains the cross-subsidy to those properties, outlines some options for reducing the risk of flooding in those areas, and briefly reviews the other cross-subsidies in the flood insurance program.

Residual Risk Behind Levees

NFIP premium rates provide a cross-subsidy to policyholders whose properties are in areas protected by levees that have been certified as capable of withstanding a 100- year flood. The cross-subsidy arises because the same Zone X rates apply to those areas as to areas that naturally lie outside 100-year floodplains, even though actual risks in the levee-protected areas tend to be higher.[2]

That difference in risks—called residual risk—occurs for two reasons. First, levees may fail to provide the expected level of protection because they subside (settle), deteriorate from tree growth, or experience other maintenance problems. The NFIP does not require periodic inspection or recertification of levees, and in its remapping program, FEMA has identified many levees that no longer comply with its regulations and should be decertified.[3] Second, even if levees perform as advertised against events up to and including a 100-year flood, the damage that results from a larger flood could be greater than in other nonfloodplain areas—for example, if levees are not just overtopped but breached.

The significance of residual risk for the NFIP is illustrated by the losses that New Orleans suffered from Hurricane Katrina, in which levee failure played a major role. Likewise, Sacramento could experience billions of dollars of losses if a major flood overtopped the levees in California's Central Valley.

Reducing Flood Risks to Areas Behind Levees

The interest of policymakers and analysts in levees and other flood-control structures focuses less on those structures' implications for the NFIP than on the adequacy of the protection they provide. The two issues are related, however. Currently, the NFIP treats areas behind levees built to protect against 100-year floods and areas that naturally lie outside 100-year floodplains the same for regulatory as well as rate-setting purposes. Thus, such levee-protected locales are not subject to the land-use restrictions that apply inside floodplains, and property owners in those areas do not have to meet the mandatory-purchase requirement that would otherwise apply to properties with federally insured or federally regulated mortgages.[4] As early as 1980, FEMA found that communities were building levees to the 100-year standard to avoid those requirements.[5] The result may be to encourage development in risky areas without encouraging mitigation measures beyond the levees themselves.

Some observers have proposed making changes to the NFIP to reduce flood damage in areas behind levees. One proposal would authorize or direct FEMA to take account of residual risk in setting premiums for structures in those areas. Besides improving the actuarial soundness of the NFIP, accounting for that residual risk in premium rates could help make property owners more aware of the limitations of the protection that levees provide. By itself, however, that approach would not affect the regulatory status of the protected areas.

A second option would be to increase the level of protection a levee must provide in order for the protected area to be exempt from the mandatory-purchase and flood-plain management requirements. Some U.S. analysts have called for the exemption standard to be increased from 100-year protection to 500-year protection, at least in urban areas.[6]

Other proposals to lessen flood damage in areas protected by levees would not involve the NFIP. An example is the call by some experts for more systematic inspection and maintenance of levees to better ensure that they will perform as designed.[7]

Other Cross-Subsidies in the Flood Insurance Program

Residual risk behind levees has drawn particular attention from policymakers in part because of the catastrophic results of levee failures in New Orleans. However, other cross-subsidies exist under the NFIP's current rate structure.

Besides areas protected by levees, another part of Zone X that faces unpriced above-average risks is 100-year flood-plains smaller than one square mile. Those areas are generally included with nonfloodplain locations in Zone X because identifying and mapping such small areas would be prohibitively expensive.[8] Another source of cross-subsidies in the NFIP is the equal treatment of Zone AE properties in inland and coastal areas. Those coastal properties tend to experience more damage for a given depth of flooding (though less than properties in Zone VE, which are exposed to waves of three feet or more during a 100-year flood), but that difference is not factored into premiums. The effect is to raise the estimates of damage as a function of depth that FEMA uses for all Zone AE structures, resulting in a cross-subsidy.

Grandfathering properties at lower rates than they would otherwise be eligible for is another type of unpriced risk—though one driven by FEMA policy rather than by legal constraints or data limitations. Most of the properties that are grandfathered after a flood map is revised are Zone X properties that would have been reclassified into an A or V zone.[9] Continuing to include those properties in the group insured at the Zone X standard rate makes the average losses in the group higher than they would be otherwise, creating a cross-subsidy from owners of "true" Zone X properties. Although such grandfathering encourages the policyholders who provide the cross-subsidy to reduce or drop their coverage (an incentive that grows stronger as the percentage of grandfathered properties increases), FEMA allows grandfathering in order to hasten the incorporation of new information into communities' floodplain management efforts by reducing property owners' opposition to the revised maps.

APPENDIX C: THE HEINZ CENTER'S STUDY OF EROSION RISKS

Coastal erosion of beaches, bluffs, and cliffs is a complex phenomenon, influenced by both natural factors and human activities. Areas subject to erosion pose two special challenges to the efforts of the Federal Emergency Management Agency (FEMA) to map flood risks. FEMA's maps reflect the fact that such areas may experience more damage from a given storm if it washes away the land beneath buildings' foundations. However, the maps do not reflect the fact that erosion moves the coastline inward, leaving properties closer to shore and hence at greater risk. For example, homes that were previously classified as outside the coastal flood zone—defined as the area exposed to damage from waves of three feet or higher in a 100-year flood—may be exposed to such waves after erosion occurs.

The Scope of the Erosion Problem

In 2000, the Heinz Center produced a study for FEMA on erosion issues in the National Flood Insurance Program (NFIP).[1] The study reached several conclusions:

- NFIP policies cover most erosion damage, but the program does not explicitly include the risk of such damage in its premium rates.[2] Moreover, it does not map existing erosion-hazard areas, so flood insurance maps may be misleading.
- On average, the risk of damage from erosion is almost equal to that from flooding along the Atlantic and Gulf coasts, where losses of roughly two to three feet of shoreline per year are common. In the next 60 years, about 25 percent of homes within 500 feet of the coast (outside major cities, which are well protected against erosion) will be lost nationwide because of erosion-related damage. That estimate translates to about 1,500 homes per year. However, not all coastal regions are subject to erosion, and loss rates vary considerably among communities.
- Over the next 60 years, erosion will cost the NFIP an average of $80 million more per year in claims, assuming no new development, changes in participation rates, or additional actions to protect

shorelines. Of that $80 million, the Atlantic coast will account for $70 million and the Gulf coast for $10 million.[3] (Losses on the Pacific and Great Lakes coasts are expected to be fairly small.)

- FEMA estimates that a nationwide program to map erosion-hazard areas would cost about $44 million— or less than $5 million a year over the 10-year useful life of erosion maps. Such mapping would be cost-effective if it discouraged even a small fraction of current development or encouraged mitigation of existing properties.
- FEMA has responded to erosion risks by raising rates in all V zones, but those uniform increases create cross-subsidies within the zones from policyholders whose properties do not face such risks.

In principle, the risk of erosion would not be a source of additional costs to the NFIP if maps and premium rates could be updated after each change in a shoreline. In practice, however, flood insurance maps are not revised that often. And even when an area is remapped, FEMA generally grandfathers the existing properties, continuing to insure them at rates that ignore any changes in their zone classification or elevation relative to the base flood level.

FEMA believes it would need authorization and funding from the Congress for erosion mapping before it could impose additional charges in erosion-hazard areas. Lacking such authorization, FEMA began a multiyear plan in May 2001 to increase rates for all policies in coastal zones above currently required levels. Boosting premiums now has two advantages for the agency: It prefunds some future erosion costs, and it provides a higher base from which future premiums can grow at the maximum statutory rate of 10 percent a year. Over the first eight years of the plan, the cumulative increase for full-risk policies in coastal zones averaged 78.2 percent.[4]

Including Erosion Risks in NFIP Premiums

Such large rate increases, and the need for continued increases in the future, heighten questions about the efficiency and fairness of the NFIP's current rate structure. Large cross-subsidies undermine mitigation incentives for policyholders who receive the subsidies and encourage policyholders who provide them to drop out of the program. If the goal were to set actuarially fair rates to protect taxpayers and to keep one class of policyholders from subsidizing another, then erosion risks would be mapped and separately priced.

The Heinz Center's study included estimates of the increases needed to price erosion risks under various approaches that draw the risk classes more or less finely. In particular, the study estimated rates for the following scenarios:

- Replace the existing coastal zones with a new coastal high-hazard zone that includes the current areas plus any areas subject to erosion within the next 60 years.
- Establish a new 60-year erosion-hazard zone, which would overlap but be distinct from the existing coastal zones. (About two-thirds of insured properties in the existing zones do not face erosion risks, according to the study.)
- Establish new erosion-hazard zones that classify the risks more finely, such as one zone for areas subject to erosion within the next 20 years and a second zone for areas expected to erode between 20 and 60 years from now.

The specific estimates from the Heinz study are no longer meaningful because they do not reflect the large rate increases that FEMA has already imposed since 2001. However, the relative magnitudes still help to illustrate the contrasts among the approaches. In particular, the rate increase for a one- to four-family home that would apply in a separate 20-year erosion zone is more than 10 times the increase that would apply if the burden of paying for erosion risk was shared evenly throughout a coastal high-hazard zone: $11.40 per $100 of coverage versus $0.90 (see Table C-1).[5]

The choice of how broadly or narrowly to define risk classes entails a trade-off. On the one hand, average pricing is simpler and distributes the burden of risk more broadly, thereby minimizing the required increases. (According to an estimate in the Heinz study, few people are likely to purchase erosion coverage voluntarily at rates above $5.00 per $100.)[6] On the other hand, pricing that is based on narrow risk classes minimizes cross-subsidies and thus avoids discouraging participation by people whose risks are lower than average. Such pricing also sends the most accurate signals of the true costs of developing or maintaining property in high-risk areas.

Besides charging full-risk insurance premiums, another way to limit erosion risks is by restricting land use. According to the Heinz study, most of the nation's 30 coastal states already impose setback requirements to prohibit new construction close to the shoreline, but the specific requirements vary widely. Seven states define the setback in terms of the area expected to be exposed to the risk of erosion within a certain period (typically 30 years for houses and 60 years for larger structures); other states define it as a certain distance (typically 25 to 100 feet) from some measure of the shoreline.[7] FEMA could set community standards that incorporate erosion risks, as it already does for flood risks. Two other approaches are to spend more money on shoreline-protection measures (such as beach replenishment, dune restoration, and shoreline hardening) or on buyouts or similar inducements to encourage existing owners to relocate.

Table C-1. Illustrative Rate Increases to Price Erosion Risks (Dollars per $100 of coverage for a 1- to 4-family residence)

	Increase in Unsubsidized Rates		Increase in Subsidized Rates	
Policy Option	For a Residence in a Coastal Flood Zone That Is Not an Erosion-Hazard Area	For a Residence in an Erosion-Hazard Area	For a Residence in a Coastal Flood Zone That Is Not an Erosion-Hazard Area	For a Residence in an Erosion-Hazard Area
Combine Coastal Flood and				
Erosion Risks into a				
Coastal High-Hazard Zone	0.90	0.90	0.35	0.35
Introduce an Erosion-Hazard				
Zone for 0- to 60-year EHAs	n.a.	2.45	n.a.	0.95
Introduce Two Erosion-Hazard				
Zones				
0- to 20-year EHAs	n.a.	11.40	n.a.	4.35
20- to 60-year EHAs	n.a.	1.75	n.a.	0.65

Source: Congressional Budget Office based on H. John Heinz III Center for Science, Economics, and the Environment, *Evaluation of Erosion Hazards* (report prepared for the Federal Emergency Management Agency, April 2000), Table 6.10, p. 181.

Notes: EHAs = erosion-hazard areas; n.a. = not applicable.

These estimates assume that the National Flood Insurance Program pays 85 percent of damages from hurricanes, that it has overhead costs of 35 percent, and that premiums for subsidized structures are about 38 percent of full-risk rates.

SELECTED BIBLIOGRAPHY

The following publications provide further information about the National Flood Insurance Program and related issues:

Blais, Neil C., and others. *Managing Future Development Conditions in the National Flood Insurance Program.* Report prepared for the Federal Emergency Management Agency. October 2006. www.fema.gov/library/viewRecord.do?id=2596.

Congressional Budget Office. *Value of Properties in the National Flood Insurance Program.* June 2007.

———. *Potential Cost Savings from the Pre-Disaster Mitigation Program.* September 2007.

Department of Homeland Security, Office of Inspector General. *FEMA's Implementation of the Flood Insurance Reform Act of 2004.* OIG-09-45. March 2009.

Dixon, Lloyd, and others. *The National Flood Insurance Program's Market Penetration Rate: Estimates and Policy Implications.* Santa Monica, Calif.: RAND Corporation. 2006. www.fema.gov/business/nfip/nfipeval.shtm.

Federal Emergency Management Agency. *Managing Floodplain Development Through the National Flood Insurance Program. Undated. www.fema.gov/library/* viewRecord.do?id=2108.

———. "Repetitive Loss Corner: Frequently Asked Questions." May 26, 2008. http://fema.nfipnextgen. com/rl_corner.html.

———. *Risk Mapping, Assessment, and Planning Multi-Year Plan: Fiscal Years 2010–2014.* March 16, 2009. www.fema.gov/library/view Record.do?id=3587.

———. "Mitigation Fact Sheets." June 4, 2009. www.fema.gov/media/fact_sheets/mitigation.shtm.

Galloway, Gerald E., and others. *Assessing the Adequacy of the National Flood Insurance Program's 1 Percent Flood Standard.* Report prepared for the Federal Emergency Management Agency. Washington, D.C.: American Institutes for Research. October 2006. www.fema.gov/library/viewRecord. do?id=2595.

Government Accountability Office. *Flood Insurance: FEMA's Rate-Setting Process Warrants Attention.* GAO-09-12. October 31, 2008.

———. *Information on Proposed Changes to the National Flood Insurance Program.* GAO-09-420R. February 27, 2009.

Hayes, Thomas L., and others, *Actuarial Rate Review*. Federal Emergency Management Agency, National Flood Insurance Program. Various years. www.fema. gov/business/nfip/actuarial_rate.shtm.

H. John Heinz III Center for Science, Economics, and the Environment. *Evaluation of Erosion Hazards*. Report prepared for the Federal Emergency Management Agency. April 2000. www.heinzcenter.org/ publications/#majorreports.

Interagency Levee Policy Review Committee. *The National Levee Challenge: Levees and the FEMA Flood Map Modernization Initiative*. September 2006.

National Research Council, Committee on FEMA Flood Maps. *Mapping the Zone: Improving Flood Map Accuracy*. Washington, D.C.: National Academies Press. 2009.

Tobin, Richard J., and Corinne Calfee. *The National Flood Insurance Program's Mandatory Purchase Requirement: Policies, Processes, and Stakeholders*. Report prepared for the Federal Emergency Management Agency. Washington, D.C.: American Institutes for Research. March 2005. www.fema.gov/business/ nfip/nfipeval. shtm.

End Notes

[1] Figures are from Federal Emergency Management Agency, *Community Status Book Report: Nation* (October 14, 2009), available at www.fema.gov/cis/nation.pdf; personal communication from Dan R. Spafford of FEMA, June 19, 2007; and Bureau Net, National Flood Insurance Program, "Policy Statistics Countrywide as of July 31, 2009," available at http://bsa.nfipstat.com/reports/ 1011.htm.

[2] Thomas L. Hayes and Dan R. Spafford, *Actuarial Rate Review: In Support of the May 1, 2008, Rate and Rule Changes* (Federal Emergency Management Agency, National Flood Insurance Program, August 2008), p. 34, available at www.fema.gov/library/ viewRecord.do?id=3430.

[3] About 8 percent of subsidized policies cover properties in four other categories: properties in communities for which flood maps have not yet been completed; properties subject to coastal flooding that were built between 1975 and 1981, when FEMA incorporated new information about wave heights and revised its new-construction standards; properties that will have flood protection when a structural project such as a levee is completed, provided that the project is currently at least 50 percent complete; and properties in areas that are protected by flood-control structures that FEMA now deems insufficient, provided that there is a repair or upgrade plan in place that meets specific standards.

[4] Hayes and Spafford, *Actuarial Rate Review,* p. 4. That estimate assumes that if full-risk premiums were charged for those policies, they would include proportionately higher allowances paid to the private insurance companies that sell and service almost all NFIP policies under FEMA's Write Your Own (WYO) program. (Of the $3.1 billion in total

premiums for policies in force as of July 31, 2009, about $0.9 billion went to WYO companies.) However, FEMA and the companies might agree not to increase the WYO allowances proportionately because there would be no proportional increase in actual sales expenses. In that case, premiums for subsidized policies would not have to rise as much to reach the full-risk level. Under the polar opposite assumption that the total allowance would remain constant, current subsidized premiums would be about 55 percent to 60 percent of their full-risk level, on average, rather than 35 percent to 40 percent. Regardless of the assumption about the allowances, the effect of subsidized rates on the funds available to the NFIP to pay claims—and hence on the program's actuarial balance—is the same.

[5] Even when premium rates are actuarially sound and incentives are undiminished, policyholders may fail to take cost-effective mitigation measures because they perceive the risk of flooding as too small or too distant in time, or they consider the measures unaffordable. For example, see Howard C. Kunreuther and others, *At War with the Weather: Managing Large-Scale Risks in a New Era of Catastrophes* (Cambridge, Mass.: MIT Press, 2009), pp. 256–261.

[6] The federal budget displays the operations of insurance programs on a cash basis. Federal loan programs, by contrast, are presented on a credit reform basis. Under credit reform, the costs of a loan program (excluding administrative costs) are measured by the discounted present value of any subsidies in the loans' interest rates and the expected defaults. For federal insurance programs, no explicit estimate of the annual subsidy from taxpayers—even on the flood insurance policies that are explicitly priced below the full-risk level—appears in the federal budget.

[7] 42 U.S.C. 4014, 104 Stat. 1388-24.

[8] Hayes and Spafford, *Actuarial Rate Review*, pp. 15–16. The federal policy fee is $13 for policies written at preferred-risk rates in low- and moderate-risk areas. For condominium policies, the fee depends on the number of units in the building. The fee is not considered part of the premium for the purpose of calculating the allowance paid to Write Your Own companies.

[9] Expanding the types of coverage sold by the NFIP could have other benefits, however. To the extent that the new types of coverage were not already provided by private insurers, they could help alert property owners to the risks they face and reduce federal spending for disaster assistance. Adding coverage for wind damage, which is part of standard homeowner's insurance, could potentially reduce claims-adjustment costs and possibly improve the NFIP's actuarial balance (if allegations that claims adjusters report some wind damage as flood damage are true). However, adding wind coverage would entail significant administrative costs. And setting actuarially sound premiums could be difficult, particularly if the policyholders who shifted from private or state wind coverage to federal coverage were disproportionately those whose properties were at higher risk.

[10] The definition of V zones reflects the fact that "a three-foot wave generally carries enough energy to break a wall panel away from a floor to which it has been nailed"; see Federal Emergency Management Agency, *Managing Floodplain Development Through the National Flood Insurance Program* (undated), p. 3-25, available at www.fema.gov/library/viewRecord.do?id=2108.

[11] Congressional Budget Office, *Value of Properties in the National Flood Insurance Program* (June 2007), p. 4.

[12] The premium income attributable to contingency loads is not directed to any separate reserve fund for catastrophic events. It simply increases the funds available to pay claims and expenses.

[13] The discounts range from 5 percent to 45 percent and are determined according to a system of points awarded for various types of risk-reduction activities. FEMA estimated that in 2006, CRS communities accounted for two-thirds of NFIP policyholders, and their average discount was 13 percent. Those figures imply that the base rate from which discounts are taken was scaled up by about 9.5 percent.

[14] If anything, the CRS program may have a small positive impact: If it induces communities to reduce their exposure to flood hazards in ways beyond those captured in FEMA's rate-setting models, claims payments by the NFIP could decline without a corresponding decrease in premium income.

[15] A structure is underinsured if it is insured for less than its full value. All else being equal, $100,000 of coverage on a $200,000 structure costs FEMA more in claims payments than does $100,000 of coverage on a $100,000 structure. The reason is that the same event generally causes similar proportions of damage to both structures, and hence, the dollar costs will be higher for the more costly structure. In the NFIP, underinsurance can affect the value of a flood claim. Damage to a single-family dwelling that is a principal residence and is insured for at least 80 percent of its full replacement cost (or to the maximum amount available under the NFIP) is valued on the basis of replacement cost. Damage to a structure that does not meet those criteria is valued on the basis of replacement cost minus physical depreciation at the time of the loss.

[16] FEMA applies the same scaling factor to all full-risk V-zone rates. Therefore, it holds the average increase below 10 percent if necessary to avoid increases in specific rates that it considers too far above 10 percent.

[17] The adjustment for inflation uses cost factors provided by FEMA. The mix of policies refers to their distribution by zone and, for Zone X, by rate class (standard or preferred-risk policies).

[18] In addition, a problem identified by the Government Accountability Office with the quality of the data that FEMA uses in deriving the damage functions could contribute to an upward bias in flood premiums. See Government Accountability Office, *Flood Insurance: FEMA's Rate-Setting Process Warrants Attention*, GAO-09-12 (October 31, 2008), pp. 14–16, 35, 43; also see Appendix A of this CBO report. In the absence of a comprehensive review of all of the data used in the NFIP, however, CBO does not infer too much from this one known problem, which could be offset (or reinforced) by other data-quality issues.

[19] Government Accountability Office, *Flood Insurance: FEMA's Rate-Setting Process Warrants Attention,* p. 18.

[20] Ibid., p. 44. FEMA also says there that "Older maps are not always outdated." Within the 68 percent figure, 21 percent of the population is covered by maps that meet or exceed two FEMA standards: the floodplain boundary standard, which is aimed at ensuring that flood maps match the topographic data on which they are based (taking those data as given), and the engineering analysis standard, which says that maps should reflect flood data and engineering analyses that are new or have been updated or validated as adequate for current use. Maps for the other 47 percent of the population satisfy at least part of either the floodplain boundary standard or the engineering analysis standard. See National Research Council, Committee on FEMA Flood Maps, *Mapping the Zone: Improving Flood Map Accuracy* (Washington, D.C.: National Academies Press, 2009), pp. 19–20. For the purposes of this report, CBO assumes that the data on which FEMA bases its flood maps and premium rates were measured and recorded accurately. However, the National Research Council study discusses many sources of uncertainty and possible error in the maps. For example, topographic data from the National Elevation Dataset of the U.S. Geological Survey, obtained from airborne and land surveys, is often used in FEMA's approximate

studies of river flooding in lower-risk areas, but the uncertainty surrounding those data is about 10 times greater than called for by FEMA's accuracy standards (ibid., pp. 30, 36).

[21] Information received from FEMA staff during a meeting on October 3, 2007, and in an e-mail dated December 22, 2008.

[22] H. John Heinz III Center for Science, Economics, and the Environment, *Evaluation of Erosion Hazards* (report prepared for the Federal Emergency Management Agency, April 2000), p. 115, available at www.heinzcenter.org/publications/#majorreports. Erosion affects the NFIP not only by hastening the obsolescence of flood maps but also by increasing the expected damage for a given storm and a given distance from structure to shoreline.

[23] Hayes and Spafford, *Actuarial Rate Review*, pp. 18, 22.

[24] Neil C. Blais and others, *Managing Future Development Conditions in the National Flood Insurance Program* (report prepared for the Federal Emergency Management Agency, October 2006), available at www.fema.gov/library/viewRecord.do?id=2596. CBO calculated the percentage changes using the estimates shown in Tables 4 through 11, pp. 33–58.

[25] Personal communication, October 24, 2007.

[26] Nathaniel L. Bindoff and others, "Observations: Oceanic Climate Change and Sea Level," Chapter 5 in *Climate Change 2007: The Physical Science Basis* (Cambridge, England: Cambridge University Press, 2007), pp. 387–388, available at www.ipcc.ch/pdf/assessment-report/ar4/wg1/ar4-wg1-chapter5.pdf.

[27] Sea level has continued to rise at a high rate since 2003 "largely due to the growing contribution of ice loss from Greenland and Antarctica"; see Katherine Richardson and others, *Synthesis Report from "Climate Change: Global Risks, Challenges and Decisions, Copenhagen 2009, 10–12 March"* (Copenhagen: University of Copenhagen, 2009), p. 10, available at http://climatecongress. ku.dk/pdf/synthesisreport.

[28] The statements by FEMA staff referred to in this paragraph are from a meeting with CBO staff on October 3, 2007, and e-mails dated December 22, 2008, and October 23, 2009.

[29] Federal Emergency Management Agency, *Risk Mapping, Assessment, and Planning Multi-Year Plan: Fiscal Years 2010–2014* (March 16, 2009), available at www.fema.gov/library/viewRecord. do?id=3587.

[30] 44 C.F.R. 65.3.

[31] U.S. Climate Change Science Program and the Subcommittee on Global Change Research, *Weather and Climate Extremes in a Changing Climate—Regions of Focus: North America, Hawaii, Caribbean, and U.S. Pacific Islands* (June 2008), available at www.climatescience.gov/Library/sap/sap3-3/final-report/ default.htm.

[32] Government Accountability Office, *FEMA's Rate-Setting Process Warrants Attention*, p. 46.

[33] U.S. Climate Change Science Program, *Weather and Climate Extremes*, pp. 46–50.

[34] For example, see R.M. Hirsch, "Flood Flows and Climate Variability and Change in the U.S.: An Exploration of the Literature, Theory, and Long-Term Flood Records" (abstract H111-01, American Geophysical Union, Fall 2008 meeting), available at www.agu.org/meetings/fm08/waisfm08.html; and William S. Logan and Laura J. Helsabeck, *Research and Applications Needs in Flood Hydrology Science: A Summary of the October 15, 2008, Workshop of the Planning Committee on Hydrologic Science* (Washington, D.C.: National Academies Press, 2009).

[35] Annual hurricane rates for three periods covering 1930 through 2006 are shown in a report by the catastrophe modeling firm Risk Management Solutions, *The 2006 RMS Expert Elicitation and Atlantic Hurricane Activity Rates Update* (white paper, November 2006), p. 3, available at www.rms.com/Publications/ 60HUActivityRates_whitepaper.pdf. Among the

scientific papers that find an increased risk of hurricanes in the near term are Lennart Bengtsson, "Enhanced Hurricane Threats," and Stanley B. Goldenberg and others, "The Recent Increase in Atlantic Hurricane Activity: Causes and Implications," both in *Science*, vol. 293, no. 5529 (July 20, 2001), pp. 440–441 and 474–479. After the 2005 season, Risk Management Solutions raised its estimates of hurricane losses across the Gulf Coast, Florida, and the Southeast by 45 percent, and its estimates of losses in the Mid-Atlantic and Northeast coastal regions by 25 percent to 30 percent, relative to those derived from averages for the 1900–2005 period. The other two leading catastrophe modeling firms—AIR Worldwide and EQECAT—also increased their estimates of losses, though by smaller amounts; see Howard C. Kunreuther and Erwann O. Michel-Kerjan, *Climate Change, Insurability of Large-Scale Disasters and the Emerging Liability Challenge,* Working Paper No. 12821 (Cambridge, Mass.: National Bureau of Economic Research, January 2007). Similarly, the companies that rate the financial strength of private insurers have raised their standards for the amount of capital that property and casualty insurers (whose policies generally cover wind damage but not flood damage) must hold against catastrophic risks.

[36] Kerry Emanuel, "Increasing Destructiveness of Tropical Cyclones Over the Past 30 Years," *Nature*, vol. 436 (August 2005), pp. 686– 688. The Power Dissipation Index is defined as the cubed estimated maximum sustained surface wind speed at six-hour intervals, summed over all tropical cyclones occurring in a given year.

[37] P.J. Webster and others, "Changes in Tropical Cyclone Number, Intensity, and Duration in a Warming Environment," *Science*, vol. 309 (September 16, 2005), pp. 1844–1846.

[38] Congressional Budget Office, *Potential Impacts of Climate Change in the United States* (May 2009).

[39] U.S. Climate Change Science Program, *Weather and Climate Extremes*, pp. 5–6, 56–61, 96.

[40] Greg J. Holland and Peter Webster, "Heightened Tropical Cyclone Activity in the North Atlantic: Natural Variability or Climate Trend?" *Philosophical Transactions of the Royal Society,* vol. 365 (1997), pp. 2695–2716.

[41] U.S. Climate Change Science Program, *Weather and Climate Extremes*, pp. 6, 110.

[42] Available data are particularly scant in the case of hurricanes that make landfall in the United States because those storms represent a small fraction of all North Atlantic hurricanes. Summarizing the state of knowledge, the U.S. Climate Change Science Program's report says, "It is very likely that the human-induced increase in greenhouse gases has contributed to the increase in sea surface temperatures in the hurricane formation regions. . . . However, a confident assessment of human influence on hurricanes will require further studies using models and observations" (*Weather and Climate Extremes*, p. 97).

[43] The Atlantic Multidecadal Oscillation is an ongoing pattern of long-duration changes in sea surface temperatures in the North Atlantic. Its significance is under debate: Compare National Oceanic and Atmospheric Administration, Atlantic Oceanographic and Meteorological Laboratory, "Frequently Asked Questions About the Atlantic Multidecadal Oscillation" (January 13, 2006), available at www.aoml.noaa.gov/phod/d2m_shift/amo_faq.php; and M.E. Mann and K.A. Emanuel, "Atlantic Hurricane Trends Linked to Climate Change," *Eos*, vol. 87, no. 24 (June 13, 2006), pp. 233–241, available at http://holocene.meteo.psu.edu/shared/ articles/MannEmanuelEos06.pdf. Climate change could have an impact on the large-scale oscillations themselves, but any such effect is highly uncertain at present.

[44] Swiss Re, *Influence of Global Warming on Tropical Cyclones, Hurricanes, and Typhoons* (Zurich: Swiss Reinsurance Company, October 2006).

[45] That estimate does not account for any accompanying rise in sea level. See William D. Nordhaus, *The Economics of Hurricanes in the United States*, Working Paper No. 12813 (Cambridge, Mass.: National Bureau of Economic Research, December 2006).

[46] Kerry Emanuel, Ragoth Sundararajan, and John Williams, "Hurricanes and Global Warming: Results from Downscaling IPCCC AR4 Simulations," *Bulletin of the American Meteorological Society*, vol. 89 (March 2008), pp. 347–367, available at ftp://texmex.mit.edu/pub/emanuel/PAPERS/Emanuel_etal_2008.pdf.

[47] The group included Fireman's Fund, Lloyd's, and Travelers insurance companies, reinsurer Swiss Re, and the Reinsurance Association of America. See H. John Heinz III Center for Science, Economics, and the Environment and Ceres, *Resilient Coasts: A Blueprint for Action* (2009), pp. 3–4, available at www.ceres.org/ Document.Doc?id=435.

[48] Lloyd's, *Climate Change: Adapt or Bust* (London: Society of Lloyd's, 2006), p. 17, available at www.lloyds.com/NR/rdonlyres/ 38782611-5ED3-4FDC-85A4-5DEAA88A2DA0/0/FINAL360climatechangereport.pdf.

[49] Of the 11 firms, six attributed the changes to the Atlantic Multidecadal Oscillation and five did not specify any causes; Government Accountability Office, *Climate Change: Financial Risks to Federal and Private Insurers in Coming Decades Are Potentially Significant*, GAO-07-285 (March 2007), p. 32.

[50] According to the Government Accountability Office, 36 percent of NFIP policies for residential structures carry the maximum coverage of $250,000. That share is above 50 percent in six coastal states and the District of Columbia and is 41.5 percent in Florida, which accounts for almost 40 percent of NFIP policies nationwide. See Government Accountability Office, *Information on Proposed Changes to the National Flood Insurance Program*, GAO-09- 420R (February 27, 2009). Many of the policyholders who carry the maximum coverage from the NFIP would benefit from higher coverage limits. Those few who carry additional private insurance above the limits—so-called excess-layer coverage—would have the opportunity to replace some or all of it with additional federal insurance, which could be expected to cost less (see Box 2 on page 8). Policyholders without supplemental insurance would be able to increase the total coverage on their property; many would presumably choose to raise their coverage up to the new limits or the value of their property, whichever was less. Increasing the limits might also affect the market for private primary (or first-dollar) flood insurance. However, many properties carrying that coverage are ineligible for NFIP insurance or are insured for $1 million or more.

[51] New coverage priced at break-even premiums would affect the program's actuarial balance in percentage terms: Relative to total premiums, any actuarial deficit (or surplus) would be smaller. For private insurers, adding new types of coverage that are not too highly correlated with existing coverage (which these might well be, especially in the case of increased coverage limits) is beneficial. Thanks to "the law of large numbers," such a change reduces the variability of companies' losses and thus the relative amount of capital they must maintain to keep the risk of bankruptcy below some acceptable threshold. For the NFIP, which does not rely on its own capital to sustain operations, the relevant figure is the annual expected loss (or gain) in dollar terms, not in percentage terms.

[52] The data analyzed in this section were provided to CBO by FEMA.

[53] The definitions of both RLPs and SRLPs require that at least two of the payments occur within 10 years of each other. The criteria for SRLPs were originally established for residential properties housing one to four families. They appear in section 102 of the Bunning-Bereuter-Blumenauer Flood Insurance Reform Act of 2004 (42 U.S.C. 4001, 118 Stat. 712), available at www.fema. gov/pdf/nfip/fira2004.pdf. FEMA later used the same criteria to

define larger residential SRLPs. Nonresidential properties are not eligible for mitigation assistance under the pilot program created by the 2004 Flood Insurance Reform Act, but the same criteria can be applied to them as well. The roughly 10,000 properties identified here as SRLPs include 8,859 residential properties and 1,195 nonresidential properties.

[54] Like other insured properties, RLPs and SRLPs may be covered by more than one policy contract. For example, a rental property may have a contract covering the structure and another covering the contents, purchased by the landlord and tenant, respectively. Moreover, for contracts bought by condominium associations, FEMA attributes one policy to each condo unit.

[55] The share of total payments attributable to current and former RLPs was closer to one-third before 2005, a year that was comparatively less catastrophic for RLPs than for NFIP properties as a whole. That year's losses account for one-quarter of total payments to RLPs since 1978 but nearly one-half of payments to all insured properties.

[56] To illustrate, the probability that a property with a flood risk of 1 percent per year will experience two or more floods in 25 years is 2.6 percent. But the probability that such a property will flood four or more times over that period is 0.01 percent (1 in 10,000). For six or more losses, the probability is less than 0.00002 percent (2 in 10 million).

[57] Even with their discounted rates, subsidized properties in high-risk A zones are subject to higher premiums, on average, than their unsubsidized counterparts are. The reason is that the subsidized properties are at much greater risk because they predate their community's entry into the NFIP, with the accompanying changes in floodplain management and building codes.

[58] Whereas the "policy" and the "contract" are the same thing for most NFIP coverage, when condominium associations purchase contracts, FEMA attributes one policy to each condo unit. Thus, the higher ratios of policies to contracts—3.5 for RLPs and 4.5 for SRLPs in V zones compared with the average of 2.2 for all full-risk policies in those zones—indicates a greater prevalence of condominium policies.

[59] See the Bunning-Bereuter-Blumenauer Flood Insurance Reform Act of 2004, 42 U.S.C. 4001, 118 Stat. 712.

[60] See Department of Homeland Security, Office of Inspector General, *FEMA's Implementation of the Flood Insurance Reform Act of 2004,* OIG-09-45 (March 2009), pp. 5, 8, 16; and Federal Emergency Management Agency, "FY 2009 SRL Allocations and Set Aside" (March 9, 2009), available at www.fema.gov/government/grant/srl/fy_2009_srl_allocations_and_set_aside.shtm.

[61] A mitigated RLP that experiences a subsequent insured flood loss is put back on the active list.

End Notes for Appendix A

[1] The very few areas for which FEMA has not yet analyzed flood hazards are labeled Zone D.

[2] FEMA uses two other labels in the A-zone category, but all of the properties in those areas are insured at subsidized rates rather than full-risk ones. Zone A99 and Zone AR refer to areas where the annual chance of flooding will be less than 1 percent upon completion of a structural flood-control project, such as a levee. Zone A99 designates an area where a new project is under way and more than half finished. Zone AR designates an area where an existing flood-control project is now considered insufficient and where the schedule and plan for making repairs or upgrades meet specific standards.

[3] Case-by-case rating is also used for properties in A or V zones that were built after the community's flood insurance rate map (FIRM) was created and that lie below the base flood elevation. Such properties are very rare because communities that participate in the NFIP must implement zoning rules and building codes that require all new (post-FIRM) structures in the identified 100-year floodplains to be built at or above the base flood elevation. Thus, post-FIRM properties that do not meet that requirement can exist only through implementation errors or enforcement problems.

[4] More precisely, the base flood is the one for which the annual probability of a flood *that large or larger* is 1 percent. Similarly, the 10-year flood is the one that has a 10 percent (1/10) chance of occurring or being exceeded in any given year.

[5] The Government Accountability Office has expressed concern that FEMA has not updated its PELV curves since the early 1980s; see Government Accountability Office, *Flood Insurance: FEMA's Rate-Setting Process Warrants Attention,* GAO-09-12 (October 31, 2008), pp. 13–14, 34. The concern is that the shape of individual PELV curves may have changed in the past 25 years. In other words, although they remain anchored at the same relative heights for the 10-year and 100-year floods, some of them may now curve more or less sharply between or outside their two anchor points than they did before. CBO believes that changes in the precise shape of the six curves are a smaller source of potential error in FEMA's rates than changes in the BFEs themselves and changes in the distributions of the 10-year flood heights relative to BFE (which determine the relative weights of the six curves in the national averages).

[6] However, FEMA uses the same damage function for inland and coastal areas in Zone AE. A given depth of flooding tends to do more damage in coastal areas (even outside V zones), so using a single average damage function for Zone AE results in a cross-subsidy. FEMA is currently studying that issue.

[7] The Government Accountability Office has also questioned the reliability of the data FEMA uses in deriving the damage functions; ibid., pp. 14–16. It found evidence that many entries in the claims database show a flood depth of zero feet (FEMA's classification for depths of a few inches) not because the adjustors who filed the claims actually reported zero feet but because they reported no depth figure at all. Misattributing damage caused by deeper flooding to shallower flooding, which occurs more frequently, should contribute an upward bias to FEMA's estimates of future claims. However, CBO does not know of any comprehensive review of the data that underpin the NFIP, and the program could have other data-quality problems that have the opposite effect.

[8] See Clayton H. Hardison and Marshall E. Jennings, "Bias in Computed Flood Risk," *Journal of the Hydraulics Division: Proceedings of the American Society of Civil Engineers,* vol. 98 (March 1972), pp. 415–427.

[9] According to FEMA staff, the need to correct for small-sample bias does not arise in Zone VE because coastal communities tend to have longer records of floods. The correction also does not apply in Zone X because, as mentioned above, rates for that zone are set on the basis of professional judgment, informed by the program's actual experience of losses, not on the basis of estimated event frequencies.

[10] See Howard Leikin and Saul Singer, "National Flood Insurance Program Rate Review Process," in *NFIP Interactive Rate Model Documentation* (June 1987), p. 3.

[11] For example, guidelines for mapping areas with short historical records specify that regional data should be averaged with local data for part of the analysis (estimating skewness). See Interagency Advisory Committee on Water Data, *Guidelines For Determining Flood Flow Frequency,* Bulletin 17B of the Hydrology Subcommittee (Department of the Interior,

March 1982). The primary source for regional data included in Bulletin 17B has not been updated since 1972, which could be causing upward or downward bias.

[12] Of the factors used in calculating the expected losses themselves, generally only the damage functions that relate water height in a structure to percentage damage change from year to year. FEMA updates those functions annually using actuarial credibility analyses of its accumulated loss experience and studies by the Army Corps of Engineers. FEMA occasionally reviews how its exposure is distributed among the six topographic profiles, on a state-bystate basis, and revises the weights as needed.

[13] In practice, the increases may average significantly less than 10 percent. As noted in the main text, FEMA interprets the 10 percent cap on average increases to mean that it should not raise the rates on particular groups of properties by much more than 10 percent. That interpretation can constrain the average increase in full-risk rates in Zone VE because the agency scales all of those rates upward by the same proportion. To illustrate, suppose the previous steps in the rate-setting process would lead to increases of 2.5 percent for 60 percent of the full-risk policies in Zone VE, 6 percent for 30 percent of the policies, and 7 percent for the other 10 percent of the policies, yielding an average increase of 4 percent. Multiplying each of the increases by a scaling factor of 2.5 to achieve an average rise of 10 percent would produce an increase of 17.5 percent on some policies. If FEMA decided that the maximum allowable increase on those policies was 14 percent, then the scaling factor would be 2 rather than 2.5, and the average increase for all policies would be 8 percent.

End Notes for Appendix B

[1] More precisely, such factors create a pure cross-subsidy, on average, over the long run but may result in a mixture of cross-subsidies and taxpayer subsidies in the short run. If the observed losses associated with a particular unpriced risk are below their long-run expected average, then experienced-based premiums do not reflect the full additional risk, and taxpayers implicitly bear some of the risk. That would be the case if, for example, experience before Hurricane Katrina did not reveal the full extent of the residual risk to areas protected by levees.

[2] The premium rates distinguish levee-protected areas in two special cases. Explicitly subsidized rates are available for properties that will be considered protected against a 100-year flood when a half-finished levee or other structure is completed. Properties in areas that were protected by structures that FEMA now deems insufficient are also eligible for subsidized coverage, provided the plan and schedule for making repairs or upgrades meet specific standards.

[3] Interagency Levee Policy Review Committee, *The National Levee Challenge: Levees and the FEMA Flood Map Modernization Initiative* (September 2006), available at www.fema.gov/library/ viewRecord.do?id=2677.

[4] Properties inside floodplains are required to carry NFIP coverage if they have a mortgage from a lender primarily regulated by a federal agency or a mortgage that is insured or purchased by a federal agency or government-sponsored enterprise such as Fannie Mae or Freddie Mac. Also, property owners who receive certain types of federal disaster aid after a flood must purchase and maintain NFIP coverage to remain eligible for such assistance after subsequent floods.

[5] Gerald E. Galloway and others, *Assessing the Adequacy of the National Flood Insurance Program's 1 Percent Flood Standard,* report prepared as part of the 2001–2006 evaluation of the National Flood Insurance Program (Washington, D.C.: American Institutes for Research, October 2006), available at www.fema. gov/library/viewRecord.do?id=2595.

[6] Interagency Levee Policy Review Committee, *The National Levee Challenge*. In the Netherlands, levees and other forms of flood protection are built to protect against 10,000-year storms in coastal areas and 1,250-year floods along major rivers. In Japan, voluntary standards call for protection against 10,000-year events in heavily populated coastal areas and 200-year events in riverside population centers; Galloway and others, *Assessing the Adequacy of the National Flood Insurance Program's 1 Percent Flood Standard*, pp. 69–71.

[7] Association of State Floodplain Managers, *Levees: The Double-Edged Sword* (April 17, 2007), available at www.floods.org/PDF/ ASFPM_Levee_Policy_Challenges_White_Paper.pdf.

[8] Personal communication, representatives of the Association of State Floodplain Managers, October 24, 2007. In addition, FEMA's rates do not account for the risk of stormwater flooding (flooding from water that has not yet entered a river, stream, or other channel). In Zone AE, stormwater flooding contributes to an actuarial shortfall because such flooding is not included in the modeled flood frequencies on which rates for that zone are based. In Zone X, the result may be a cross-subsidy to owners of properties that are subject to stormwater flooding if they face higher risks overall—for example, if their risks of regular flooding are average and the stormwater risks are on top of that average risk.

[9] As noted in the main text, however, some grandfathered properties are A-zone properties that would have moved to a V zone, or A- or V-zone properties that would have remained in the same zone but been reclassified at a lower elevation. Because the "excess" flooding experienced by such properties does not feed back into higher rates for all properties in those zones—as it does for grandfathered properties in Zone X—the result is an implicit subsidy from taxpayers rather than a cross-subsidy.

End Notes for Appendix C

[1] H. John Heinz III Center for Science, Economics, and the Environment, *Evaluation of Erosion Hazards* (report prepared for the Federal Emergency Management Agency, April 2000), available at www.heinzcenter.org/publications/#majorreports.

[2] The program's definition of a flood includes "the collapse or subsidence of land . . . as a result of erosion or undermining caused by waves or currents of water exceeding anticipated cyclical levels or suddenly caused by an unusually high water level in a natural body of water"; ibid., p. 45. Excluding wet-weather erosion from NFIP coverage would be impractical because it is usually impossible to distinguish between losses from erosion and losses from flooding in such cases. In contrast, dry-weather collapse of a building because of erosion is not covered under the NFIP.

[3] Ibid., Table S.2, p. xxix. Surveys indicate that only about half of the houses in high-erosion areas on the Atlantic and Gulf coasts have flood insurance.

[4] In May 2008, rates for both full-risk and subsidized coastal-zone policies grew by roughly the maximum of 10 percent. Thomas L. Hayes and Dan R. Spafford, *Actuarial Rate Review: In Support of the May 1, 2008, Rate and Rule Changes* (Federal Emergency Management Agency, National Flood Insurance Program, August 2008), pp. 18 and 22, available at www.fema.gov/library/ viewRecord.do?id=3430.

[5] If the Congress chose to maintain the current proportional subsidies for older properties (primarily ones built before 1975 or before the local community joined the NFIP), those increases in premiums would range from \$0.35 to \$4.35 per \$100 of coverage. CBO estimates that about half of the policies in coastal zones are explicitly subsidized (that estimate is based on Hayes and Spafford, *Actuarial Rate Review*, p. 34). As originally defined by the Heinz Center, the scenario involving two erosion-hazard zones applied the erosion premiums only to new structures (whose owners could decide to build outside the 0- to 20-year erosion zone to avoid the higher premiums there); older subsidized properties would not be affected. The figures used here are for one- to four-family homes. Other types of properties would face analogous increases.

[6] Heinz Center, *Evaluation of Erosion Hazards*, p. 160.

[7] Ibid., p. xxxviii.

In: Hurricane Damage
Editor: Martha Atkinson
ISBN: 978-1-53610-035-8

Chapter 4

THE MACROECONOMIC AND BUDGETARY EFFECTS OF HURRICANES KATRINA AND RITA: AN UPDATE*

Congressional Budget Office

CONGRESSIONAL BUDGET OFFICE
U.S. Congress
Washington, DC 20515

Douglas Holtz-Eakin, Director

September 29, 2005

Honorable Jim Nussle Chairman
Committee on the Budget
U.S. House of Representatives
Washington, DC 20515

Dear Mr. Chairman:

As you requested in your letter of September 23, 2005, the Congressional Budget Office (CBO) has reviewed its estimates of the likely consequences of Hurricanes Katrina and Rita for the national economy, federal receipts, and

* This is an edited, reformatted and augmented version of a report issued by the Congressional Budget Office, September 29, 2005.

federal outlays. The preeminent cost of the disasters is the tragic loss of life. However, both hurricanes have had economic impacts, some of which will probably spread throughout the economy. Those impacts arise from the disruption of production (especially of oil, oil products, and natural gas), from spending in the affected areas, and from the loss of wealth of those most directly affected.

However, even with the additional impacts of Hurricane Rita, the macroeconomic consequences of the hurricanes appear more modest than those that CBO estimated for Hurricane Katrina in its letter to Senator Frist on September 6, 2005. According to CBO's estimates, while gross domestic product (GDP) will grow about half a percentage point more slowly over the second half of 2005 as a result of the hurricanes, most of that slowing will occur in the third quarter. CBO anticipates that, with private and government support for recovery and rebuilding, GDP growth will not be much affected in the fourth quarter of 2005 and could even be somewhat higher than was projected before the hurricanes. Overall, GDP will return to its previous trend by early 2006, CBO projects, and subsequently rise above that trend as rebuilding raises overall economic activity. Employment growth will roughly mirror the impacts on GDP. The pace at which consumer energy prices return to their pre-Katrina levels will be the dominant factor in determining the duration of higher inflation.

But those estimates are necessarily uncertain. CBO, like other forecasters, bases its analyses on available information about the losses in the stricken region and about the flow of resources to the region. New information on those matters, as well as about the damage to and recovery of energy resources, will emerge in coming weeks, and estimates of economic effects will change as that information becomes available.

The disasters also will have a significant impact on the federal budget. Because the overall economic impacts will probably be offset and then reversed by the end of fiscal 2006, the budgetary effects will be largely the result of legislation. At this point, the scope of such legislation and its ultimate costs are yet to be determined. The President has already requested and the Congress has appropriated $62.3 billion in emergency assistance, which will raise federal outlays by roughly $30 billion in 2006 and by smaller amounts in subsequent fiscal years. Spending to pay flood insurance claims is expected to rise substantially in fiscal year 2006, and additional funding for that program is likely to be necessary. In addition, tax relief legislation enacted on September 23 is expected to reduce federal receipts by $6 billion, almost entirely in fiscal years 2006 and 2007. Finally, there has been a wide variety of legislative initiatives, as well as some discussion of budgetary changes to offset—in whole or in part—any additional federal costs.

The details of CBO's economic and budgetary analysis are in the attachment. I hope that you find the analysis useful. As you requested in your letter, CBO will provide additional updates as more information becomes available. CBO would be pleased to address any further questions that you might have.

Sincerely,
Douglas Holtz-Eakin
Director

Attachment
Identical letter sent to the Honorable Judd Gregg
cc: Honorable John M. Spratt, Jr.

SUMMARY

Hurricanes Katrina and Rita have temporarily reduced the growth of economic output, but the effects that rebuilding will have on economic activity may more than offset the drag by early next year. For the federal government, unless the Congress takes action to reduce other spending or raise taxes, the hurricanes will push the budget further in deficit for the next few years, largely because of the $62 billion appropriated for emergency assistance, but also because of various temporary changes to tax rules and a slightly weaker economy during the second half of this year.

The majority of the economic and budgetary effects of the hurricanes can be attributed to Katrina. Initial indications are that Rita had a smaller effect on nonenergy infrastructure, although its impact on oil, natural gas, and electricity resources may be greater than that of Katrina. An assessment of Rita's damage to offshore platforms and pipelines is incomplete.

The major uncertainties affecting the speed of recovery are how quickly insurance and government payments are distributed; how quickly consumer energy prices decline; and how quickly rebuilding can start, in New Orleans and elsewhere. For example, if, during 2005, about half of the private insurance claims are paid out, federal spending for relief and recovery totals about $10 billion (in transfers and outlays for goods and services), gasoline prices fall back to only about 10 percent higher than their pre-Katrina levels, and rebuilding is only slightly delayed relative to the timing experienced in previous hurricanes, then economic dislocation caused by the hurricanes probably will be offset by the reconstruction effort by early next year.

The direct, immediate reduction in employment in the areas affected by Hurricane Katrina now appears to be between 280,000 and 400,000 jobs. The effects of Hurricane Rita appear to have been considerably smaller. Much of those losses will be trimmed as some people return to their former jobs or find new jobs (including in reconstruction work) in the next few months. By early next year, the pace of reconstruction will probably cause the net effect of the hurricanes on jobs nationwide to be minimal. If, as appears likely, output bounces back to equal or exceed its previous trend by early next year, total employment will be similar to what it would have been if the hurricane had not occurred, even though some of the people who lost jobs may remain unemployed for some time.

The growth of consumer prices will be higher for the second half of this year than previously expected, primarily because of the increase in energy prices. But consumer price inflation should revert to pre-Katrina rates in the first half of 2006 if, as is generally expected, energy prices ease part of the way back to their pre-Katrina levels. Higher prices of construction materials and higher energy prices (which raise transportation costs) will tend to temporarily increase inflation in the prices for many non-energy-related goods as well as for airline, bus, and railroad fares.

EFFECTS ON GROSS DOMESTIC PRODUCT

The initial effects of the hurricanes on economic output comes from the lost production in the affected regions and the temporary spike in energy costs. However, looking forward, the impact of the hurricanes on the pace of production and income depends on what happens to four major categories of spending: investment (in business plant and equipment, commercial structures, and housing), spending on consumer durables, government spending for goods and services, and other household consumption expenditures (see Table 1).

The Congressional Budget Office (CBO) estimates that the hurricanes could reduce real (inflation-adjusted) growth of gross domestic product (GDP) in the third quarter of 2005 by between 1 and 1Y2 percentage points, but as the cleanup and repair begin, the economy is likely to grow in the fourth quarter at a rate not much different from what would have occurred without the hurricanes, and possibly even a little higher. GDP growth for the second half of 2005 is likely to be about Y2 a percentage point lower than it would otherwise have been. The spending to repair or replace capital stock (homes, business structures, and equipment) is likely to drive the level of output back to its previous trend early in 2006 and to continue to add slightly to growth during the rest of the year.

Initial Production Losses in the Directly Affected Areas

Estimates of the initial production losses in the directly affected areas are necessarily very rough. CBO estimates the direct losses of marketable goods and services to be about $28 billion to $44 billion (at annual rates) in both the third and fourth quarters. Those estimates are for initial production losses and do not take into account the effects of higher energy prices on nationwide economic activity, increased government spending, or the impact of private rebuilding.

The vast majority—$18 billion to $28 billion in each quarter—of the losses in production comes from lost oil and gas production and from refinery damage. The loss in crude oil production is expected to be 0.35 million barrels per day (mmbd) in the third quarter and 0.50 mmbd in the fourth. For natural gas, output appears to be 0.25 trillion cubic feet (tcf) lower in the third quarter and 0.30 tcf lower in the fourth. Refinery output (primarily of gasoline) is expected to fall by 0.30 mmbd in the third quarter and 0.15 mmbd in the fourth quarter. (To gauge the value of lost output, CBO multiplied those physical losses for each category by estimates of what the prices of the products would have been in the absence of the hurricane.)

Other production (excluding housing services and agriculture) may have fallen by $8 billion to $12 billion in the third quarter of 2005 and may fall by a similar amount in the fourth quarter. (Those estimates exclude rebuilding activity.) CBO's estimates are based on the number of employees displaced by the hurricanes, on the productivity of workers in that area relative to the national average (inferred from relative wage rates), and on national productivity estimates. (Estimated impacts on employment are discussed more fully below).

Table 1. Estimated Net Effect of Hurricanes Katrina and Rita on Real GDP (Billions of 2005 dollars at annual rates)

	2005	**2006**		**2007**	
	2nd Half	**1st Half**	**2nd Half**	**1st Half**	**2nd Half**
Energy Production	-18 to -28	-8 to -10	-5 to -7	-5 to -7	-5 to -7
Housing Services	-1 to -2	-2 to -4	-1 to -3	0 to -2	0 to -2
Agricultural Production	-1 to -2	0	0	0	0
Replacement Investment	7 to 11	29 to 43	30 to 46	30 to 46	25 to 37
Government Spending on Goods and Services	6 to 10	12 to 18	14 to 20	10 to 16	7 to 11
Effect of Higher Energy Prices on Nonenergy Consumption	-6 to -10	-5 to -7	-2 to -5	-1 to -3	0 to -2
Other Consumption	-8 to -12	-2 to -4	-1 to -3	-1 to -3	0 to -2
Real GDP	**-21 to -33**	**24 to 36**	**35 to 48**	**33 to 47**	**27 to 35**

Source: Congressional Budget Office.

The consumption of housing services in the region may have fallen by at most a few billion dollars. The hurricanes made uninhabitable about 0.2 percent of the nation's housing stock (owner-occupied, rental, and hotel units). Nationally, the housing stock provides housing services that the Bureau of Economic Analysis estimates were worth about $1,275 billion (at an annual rate) in the second quarter of 2005. On that basis, the loss of housing services from the hurricanes could be about $3 billion in the fourth quarter of 2005. Only one month of the third quarter—September—is affected, so the production loss in that quarter would be correspondingly smaller. Some of the apparent losses may be offset, as evidenced by the fact that the hotels and motels in Louisiana, Texas, and Mississippi that are still operating have extremely high occupancy rates and the vacancy rates of rental units are lower than they otherwise would have been.

Agricultural losses, according to the estimates of the U.S. Department of Agriculture, could amount to $0.9 billion. That could subtract between $1 billion to $2 billion, at an annual rate, from production in the second half of 2005.

Expected Impacts on Production and Incomes

In the second half of 2005, the effect of the hurricanes on overall real GDP growth will probably be modestly negative, approximately 1/2 percentage point at an annual rate, primarily because of the drop in production in the third quarter. By the fourth quarter, cleanup and repair efforts are likely to begin to spur economic activity. By the early part of 2006, the level of real GDP is likely to be at least as high as it would have been without the hurricanes, and the growth of real GDP and incomes will be modestly boosted for a year or so thereafter.

The possibility that the hurricanes could boost GDP does not mean that they are not damaging. As a measure of economic activity, GDP does not record the losses of lives and assets.

CBO's analysis of likely effects on production and incomes is based on what may happen to investment (in business plant and equipment, commercial structures, and housing), to spending on consumer durables, to government spending for goods and services, and to other household consumption expenditures. The analysis does not include any dynamic feedback effects—that is, the tendency of increased spending in one area to increase incomes and thus spending everywhere. Such effects are likely to be small, particularly if the Federal Reserve Board does not alter its apparent plan to increase its target interest rates. (The Federal Reserve increased rates by 25 basis points on September 20, as had been expected before the hurricane.)

Replacement of Lost Plant, Equipment, Residences, and Consumer Durables

Estimates of damages remain quite uncertain because the damage has not been completely surveyed. However, some private insurance sources have made initial estimates. For example, figures from Risk Management Solutions suggest that total losses—insured and uninsured—approach $140 billion.[1] (See Box 1, for a discussion of insurance payments.) Some portion of those losses will reflect business interruptions, rather than actual capital losses, so for purposes of illustration, CBO assumes a round number of $100 billion in losses to fixed capital and consumer durables.

Under an assumption that the hurricanes damaged a representative sample of the national stocks of fixed capital and consumer durables, about 40 percent of the damage would have been in housing, a little over 30 percent in business plant and equipment, nearly 20 percent in public infrastructure (roads, bridges, sewer systems, and so forth), and 10 percent in consumer durables (vehicles, appliances, furniture, and so forth) (see Table 2). As time goes on, it may be possible to base estimates on the actual amount of damage experienced in the stricken areas, but for the time being, such estimates are not available.

CBO's analysis suggests that the pace of rebuilding and replacement of lost plant, equipment, housing, and consumer durables (that is, total private replacement and rebuilding) could rise quite rapidly, to an annual rate of between $29 billion and $43 billion (in 2005 dollars) by the first half of 2006. By far, the majority of such spending—perhaps three-fourths of the total—will be in residential construction.

Table 2. Stocks of Fixed Capital in the United States, End of 2004

	Billions of dollars	**Percentage of fixed assets and consumer durables**	**Percentage of private fixed assets**
Fixed Assets and Consumer Durables Fixed assets	37,610 34,032	100.0 90.5	
Private	27,043	71.9	100.0
Nonresidential	12,569	33.4	46.5
Equipment and software	4,616	12.3	17.1
Structures	7,953	21.1	29.4
Residential	14,473	38.5	53.5
Government	6,990	18.6	
Consumer durable goods	3,578	9.5	

Sources: Congressional Budget Office; U.S. Department of Commerce, Bureau of Economic Analysis.

Government Spending

Various levels of government are spending large amounts for rescue and recovery. The federal government may have outlays of about $2 billion more in September than it otherwise would have had and perhaps $10 billion more in the fourth quarter. Spending for the next few years will also be higher as a result of the emergency appropriations and, possibly, additional funding yet to come. Some of the federal spending to date (to provide supplies and medical care, cover expenditures for the National Guard, repair levees and pumping stations, pump out New Orleans, and purchase mobile and manufactured homes, among other things) is for goods and services, but some consists of transfers to individuals and grants to state and local governments.

State and local government spending will also be higher during the fourth quarter of 2005 and in 2006 as infrastructure is rebuilt. Experience with previous disasters does not provide clear guidance as to how large the increase will be, but it may be on the order of a few billion dollars in the fourth quarter.

Consumer Spending

The hurricanes have reduced the consumption expenditures of people in the directly affected areas and for the nation as a whole through the effect of higher energy prices.

Consumption of Nondurables by Affected Residents

People who lost houses, businesses, or jobs in the hurricanes will probably, even with insurance and government assistance, face a loss of wealth that will reduce their consumption spending. In addition, at least in the short term, many residents in the stricken areas may find themselves short of cash and, without access to documentation, unable to borrow. Under those circumstances, their consumption may drop until they can get either jobs or assistance from the government or charities.

Insurance payments may be an important offset to losses of wealth, though some may be delayed by disputes over the source of the loss (as described in Box 1).

Effects of Higher Energy Prices on Nonenergy Consumption

Consumer purchases of nonenergy items nationwide will also be affected by the increases in energy prices. Such increases redistribute income from consumers to stockholders of companies producing and refining energy, some of whom are also U.S. consumers and some of whom are foreigners. If all of those parties spent about the same portion of their income on U.S.-produced nonenergy goods and services as do other consumers, there would be no net effect on nonenergy consumption. However, foreigners are likely to spend less than U.S. residents on U.S. goods, and some of those consumers facing higher energy prices may have difficulty maintaining their consumption of other goods and services as they pay more for energy. Thus, CBO assumes that about 40 percent of consumers will have to cut back on nonenergy purchases.

Box 1. Insurance Payments Associated with Hurricanes Katrina and Rita

According to Standard and Poor's, privately insured losses from Hurricane Katrina could range from $35 billion to $60 billion. Hurricane Rita, according to Risk Management Solutions, resulted in privately insured losses of $4 billion to $7 billion. By comparison, insurers paid about $32.5 billion after the September 11, 2001, attacks and just under $23 billion after last year's four hurricanes. Developing credible loss estimates for Hurricanes Katrina and Rita is difficult because of the nature of the losses. In general, private insurers are responsible only for wind losses, not losses caused by flooding, so disputes over settlements are likely. And insurers' access to the properties has been delayed by the flooding. Moreover, the extent of business interruption losses, which could be substantial, depends on the duration of the interruption, which is not yet known. (Business interruption losses were about $11 billion, or one-third of the insured losses after 9/11.) Finally, liability claims are possible from environmental damage.

Some insurance claims have already been settled, and many policyholders have received payments for temporary housing assistance. There were about 2 million claims ($22.9 billion) after last year's hurricanes, and over 95 percent were settled within 41/2 months, according to industry sources. More claims are expected from Katrina and Rita, and the settlements are likely to be much higher. Disputes over the source of the, however, could delay settlements of claims, especially if those disagreements end up in the courts. The Mississippi Attorney General filed suit on September 15 to force insurers to cover flood losses where the policyholders did not have federal flood insurance. Reinsurers might dispute obligations to primary insurers on any flood-related claims should that suit be successful. The suit could have the unintended effect of restricting the supply of private insurance in Mississippi in the future and thereby slow the pace and extent of rebuilding.

Although no estimates have been published, federal flood insurance payments are also likely to be substantial. Although a significant percentage of affected households did not have flood insurance, the National Flood Insurance Program (NFIP) has received 150,000 claims arising from Katrina to date, and NFIP has tried to speed up payments through the use of aerial photos.

EMPLOYMENT EFFECTS

Between 280,000 and 400,000 people lost jobs directly because of Hurricane Katrina. The effects of Hurricane Rita appear to have been considerably smaller, although the data are not complete. The impacts of the hurricanes on general economic activity mean that employment will be temporarily depressed—on average as well as in the stricken areas. However, the rebuilding efforts appear likely to fully offset all of the reduction in employment in the first half of 2006.

Measuring the effects of the hurricanes on employment will remain difficult, even after the Bureau of Labor Statistics begins to publish data for September. (Those data will be published on October 7 for the national level and on October 21 for the state level). The bureau will probably experience unusual difficulty in measuring employment in the storm-damaged areas (see Box 2).

Direct Effects of Katrina

Two primary sources of information are available to estimate the possible impacts of the hurricanes on employment. A lower bound can be estimated from the number of storm-related claims for unemployment insurance filed to date. The Labor Department estimates that a total of 279,000 such claims had been filed by September 24. The number could go higher. (One potential source of future claims is workers who have so far remained on the payroll even if they cannot work but who will be dropped at some point.)

The other source of information is the Bureau of Labor Statistics' report of annual average data for 2004, based on the Quarterly Census of Employment and Wages. That report includes the number of establishments, total employment, and total wages in areas affected by Katrina.[2] The figures define the jobs potentially at risk because of flooding and other damage caused by Katrina and as such represent an upper bound of the storm's effect. At the broadest level, the Federal Emergency Management Agency (FEMA), as of September 8, had designated 169 counties or parishes in four states as eligible for some form of disaster assistance; employment in those areas totaled about 5.8 million jobs, or 4.5 percent of the U.S. total. Residents of those counties might have missed at least one day of work. In the 86 counties or parishes classified as "most affected" (meaning that residents are eligible for both individual and public disaster assistance), where people could have missed a week or more of work, employment was 2.4 million jobs (or 1.9 percent of the national total). The wage bill in those counties was $76.7 billion, or 1.5 percent of the national total.

Workers in the areas that FEMA has identified as flooded and storm-damaged are those most likely to experience an extended absence from work (or even to lose their jobs permanently). According to the Bureau of Labor Statistics, in the fourth quarter of 2004 there were about 22,500 business establishments within those areas employing roughly 373,000 workers and paying $3.49 billion in wages and salaries during the quarter (not annualized) (see Table 3). Most of the at-risk employment was in flooded areas of Louisiana, while in Mississippi, virtually all of the potential employment losses were attributable to damage rather than flooding. In addition, about 265,000 workers were employed within half a mile of the storm-damaged areas in Louisiana and Mississippi—184,000 of them in Louisiana.

Direct Effects of Rita

Hurricane Rita's impacts on employment appear to have been considerably smaller than those of Katrina. While detailed data on areas damaged by Rita are not yet available, data from the Quarterly Census of Employment and Wages indicate that in the 17 counties or parishes designated as eligible for individual disaster assistance, total employment in 2004 stood at about 482,000 jobs, or 0.37 percent of the U.S. total. The wage bill in those areas was $15.7 billion, or 0.3 percent of the U.S. total.

As in the case of Katrina, actual employment losses are likely to be considerably smaller than the number of jobs potentially at risk. The timing of the storm was such that the September survey was already completed before the hurricane struck, so no measured effect will be present in the data for September. If most residents are able to return home and cleanup can begin once floodwaters have receded and power and other basic services are restored, then the impact on measured employment in October should be small. The evacuation of more than 2 million residents from the Houston metropolitan area (most of whom did not reside in counties eligible for individual disaster assistance) probably resulted in a loss of a few days' pay for some workers and reduced profits for employers who continued to pay their workers, but will not show up in the October employment data. Renewed flooding in portions of New Orleans and St. Bernard Parish might slightly delay the recovery from job losses attributable to Katrina, though it should have no impact on employment totals by the end of the year.

Box 2. Employment Measurement Issues

The Bureau of Labor Statistics recognizes that it faces unusual challenges in collecting employment data in the areas damaged by Katrina. As a consequence, some procedures will be modified. In particular, at the establishment level, nonrespondents in the most heavily affected areas will be assumed to have an employment level of zero, instead of the usual assumption made for nonrespondents that over-the-month changes were the same as at those of establishments with similar characteristics. That modification runs the risk of overstating employment losses attributable to Katrina, but the normal procedure would carry a strong risk of understating such losses. For the many multiestablishment firms that report all of their data from a single central site, it should be possible to obtain accurate data on workplaces within the disaster areas. The bureau does not intend to quantify the direct effect of Hurricane Katrina, "because its effects can not be separated from other influences on the economy, particularly at the national level."

Evacuees staying in shelters, hotels, and motels were not interviewed for the September survey of households, and there was no attempt to contact sample households in Orleans and Jefferson parishes. However, outside those parishes, people in sample households who had evacuated but returned to their homes by the week of September 19 were interviewed as usual.

Hurricane Rita probably will not affect the data for September and is likely to pose, at most, minor difficulties for the collection of establishment-level data. However, the collection of household data in Southwest Louisiana and Southeast Texas (including the Houston metropolitan area) might have been hampered by the evacuation for Rita.

Employment Pattern through Early 2006

The pattern of employment effects for the nation will tend to track the pattern for GDP discussed above. The net effect of the negative macroeconomic effects and the positive rebuilding effects will probably result in a higher level of employment by the second quarter of 2006 than would have occurred in the absence of the hurricanes. By the end of this year, the net employment loss may be less than 200,000 jobs, and by the second quarter of next year, net employment may be higher. Of course, many of the people who lost jobs because of Katrina may be unemployed for many months.

Table 3. Employment and Wages Within Areas Damaged by Hurricane Katrina, Fourth Quarter of 2004

	Alabama	Louisiana	Mississippi	Total
	Employment (Thousands of jobs)			
Flooded Area	*	305	1	306
Nonflooded Area by Degree of Damage[a]				
Limited	2	8	11	20
Moderate	*	2	6	9
Extensive	0	1	2	3
Catastrophic	0	*	35	36
Total	2	316	55	373
Wages (Millions of dollars, quarterly rate)				
Flooded Area	1	2,966	5	2,972
Nonflooded Area by Degree of Damage[a]				
Limited	16	60	111	187
Moderate	3	13	42	57
Extensive	0	4	15	18
Catastrophic	0	5	248	252
Total	20	3,047	419	3,486

Sources: Congressional Budget Office; Department of Labor, Bureau of Labor Statistics; Department of Homeland Security, Federal Emergency Management Agency.

Notes: Zones that were both flooded and storm-damaged were classified as "flooded areas." * = Fewer than 500,000.

[a]. Based on FEMA's storm-damage categories.

INFLATION

The effect of the hurricanes on inflation depends almost entirely on the path of energy prices. (A detailed discussion of factors affecting energy prices begins on page 19.) Shortages of some other commodities and increased demand for workers and materials for rebuilding will have a relatively small effect on inflation.

Gasoline prices, as compiled by the Department of Energy's Energy Information Administration (EIA), jumped sharply in late August and the initial weeks of September because of the reduction in refinery operations, but forecasts of gasoline prices remain significantly higher through mid-2006 (see Figure 1). The longer-run effect on prices stems from the prospect of reduced refinery capacity for many months and from the continued pressure of high world oil demand on crude oil prices.

Natural gas prices have also jumped in commodity markets. Although natural gas prices for homeowners will not rise as much in percentage terms, they may still be almost 40 percent higher this winter than last (see Figure 2). The persistence of the effect of the hurricanes on natural gas prices may be even greater than that for gasoline. Utilities' demand for natural gas is expected to remain high even though prices are high.

Electricity prices for residential consumers will ultimately increase as a result of the hurricanes, although not immediately because of the regulation of electricity prices for those users. In the future, higher natural gas prices and the cost of rebuilding capital stock damaged by the storms will lead to higher electricity prices.

Retail food prices are expected to be slightly higher because of increased transportation costs, even though the outlook for farm prices is slightly lower. Farm prices are lower in part because of hurricane-related transportation delays, but primarily because agricultural production estimates for most crops increased recently for reasons unrelated to the hurricanes.

The direct short-term effects of the hurricanes on the rise in the consumer price index for all urban consumers (CPI-U) stemming from the increase in energy prices will be large. From those direct effects alone, the growth of the CPI-U between the fourth quarter of 2004 and the fourth quarter of 2005 may be almost 1 percentage point higher than would have occurred in the absence of the hurricanes. During 2006, however, CPI-U inflation may be slightly lower than previously anticipated, as the effect of the hurricanes on energy prices dissipates.

Effect on the Federal Budget

It is still unclear what the ultimate impact of Hurricanes Katrina and Rita will be on the federal budget. Much will depend on policy decisions yet to be made as to the scope of the federal government's role.

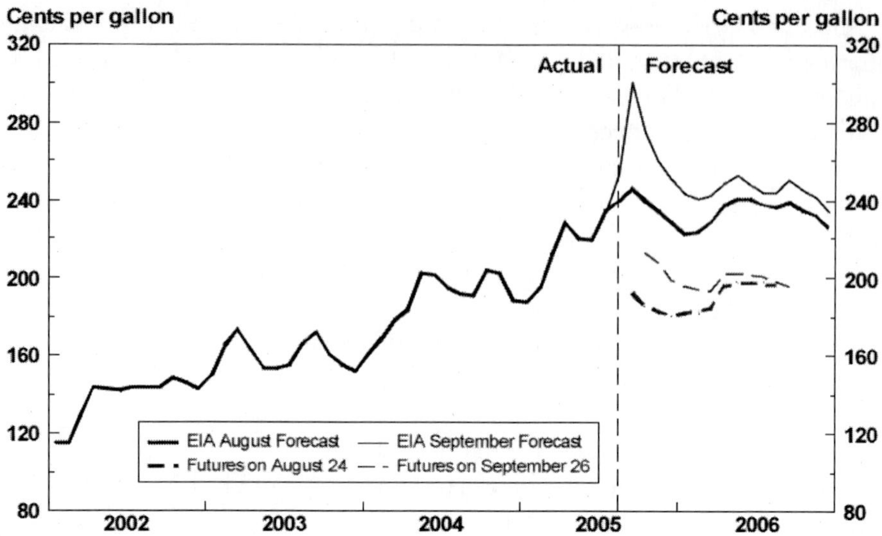

NOTE: EIA Forecasts are for the retail price of gasoline, whereas the futures are for a wholesale price.

Figure 1. Change in Gasoline Price Forecasts: August Forecast vs. September Forecast.

Federal Spending

To date, the President has requested and the Congress has appropriated \$62.3 billion in emergency assistance in response to Katrina. Almost all of that amount—\$60 billion—was provided to the Federal Emergency Management Agency's disaster relief account; some of those funds could be used for Hurricane Rita or other disasters, if necessary. As of September 21, FEMA had obligated about \$12 billion of that appropriation, and allocated another \$4 billion for future obligation for activities related to Hurricane Katrina. Of that \$16 billion, \$6.3 billion has been allocated for housing assistance and the acquisition of manufactured housing, \$3.1 billion has been committed to states in the form of goods and services for relief activities, and \$3.5 billion will be used to reimburse billion has been allocated for housing assistance and the acquisition of manufactured housing, \$3.1 billion has been committed to states in the form of goods and services for relief activities, and \$3.5 billion will be used to reimburse other federal agencies—particularly the Army Corps of Engineers and the Defense Department (DoD)—for their disaster relief efforts. In addition, those agencies have received funding of their own—\$400 million for the Corps and \$1.9 billion for DoD (for costs associated with the deployment of military personnel in support of relief efforts, for the evacuation of military personnel and their families, and for short-term repairs to military facilities).

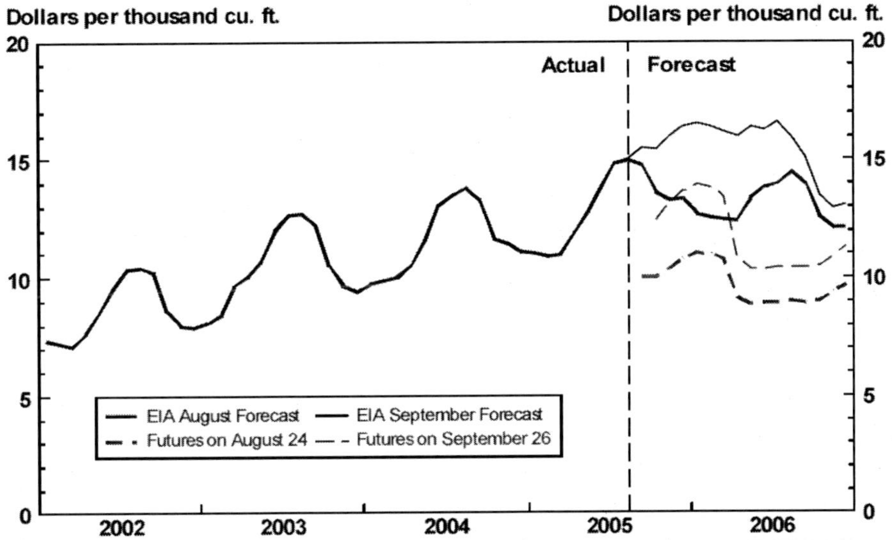

Note: EIA Forecasts are for the price of natural gas for residential consumers, whereas the futures are for the Henry Hub price.

Figure 2. Change in Natural Gas Price Forecasts: August Forecast vs. September Forecast.

CBO anticipates that outlays from the supplemental appropriations will be about $2 billion in September, about $30 billion in fiscal year 2006, and the remainder over the following three years. Although billions of dollars may be obligated in this month (that is, within fiscal year 2005), most of the checks are likely to be written in subsequent months. Any substantial spending on reconstruction activities will occur over a period of several years.

In addition to supplemental appropriations for disaster relief, the Congress and the President have enacted a $2 billion increase in borrowing authority for the flood insurance program, bringing the total available borrowing authority to $3.5 billion. CBO expects that payments for flood insurance claims resulting from the hurricanes will reach $1 billion to $2 billion in the first quarter of fiscal year 2006. Historically, FEMA has paid at least one-third of the total value of flood insurance claims within three months of an event, and 90 percent of those claims within one year. Given the magnitude of Hurricane Katrina, CBO expects that the initial rate of payments, within the first three months, will be slower. However, because FEMA is already implementing expedited procedures for paying claims in affected areas, the percentage of claims paid by the end of fiscal year 2006 will probably be in line with historical experience. CBO estimates that additional outlays in 2006 from currently available authority will total between $3 billion and $4 billion.

Although CBO does not have sufficient information at this time to estimate the total value of claims that FEMA is likely to face because of the recent hurricanes, information from the agency about the amount of flood insurance in force in the affected areas suggests that those losses will significantly exceed the $3.5 billion that CBO estimates is currently available to pay such claims.

The Congress and the President also enacted the TANF Emergency Response and Recovery Act of 2005 (Public Law 109-68), which provides additional funds to states that were damaged by Hurricane Katrina or those that are hosting evacuees from the hurricane to provide benefits to needy people. That legislation will cost about $400 million, CBO estimates, mostly in 2006.

Some federal programs may experience temporary increases in spending as evacuees become eligible for benefits (such as Medicaid) that they would not normally collect. Also, workers who are now unemployed as a result of the hurricanes may receive unemployment benefits, which CBO estimates could cost about $600 million (in addition to disaster unemployment assistance, which is paid by FEMA). The government is likely to lose at least a few hundred million dollars in royalty payments for oil and gas leases on the Outer Continental Shelf; on the other hand, the government will receive about $700 million from the sale of 11 million barrels of oil from the Strategic Petroleum Reserve.

Legislation passed or under consideration in the House or Senate would, if enacted, further add to the federal deficit. For example, the Senate has added $4.3 billion to the Commerce-Justice-Science appropriation bill, mostly for emergency housing assistance. CBO estimates that $2.5 billion of that sum would be spent in 2006.

Federal Revenues

The hurricanes will affect total federal receipts through their impact on the overall economy and through delays in the payments of taxes and other tax relief. The temporary reduction in economic growth will reduce receipts but not significantly. A reduction in overall economic growth of 0.5 percentage point for a full year might reduce receipts by between $5 billion and $10 billion, or less than a billion dollars per month on average. But the effect of the hurricane in reducing economic growth is expected to be of much shorter duration, and the changes in receipts are not likely to stand out from total federal receipts that average about $175 billion per month. As the economy picks up with recovery and reconstruction, the negative effect on overall receipts should dissipate.

Delays in the payment of taxes could affect the timing of several billion dollars or more in receipts over coming months. As a result of Internal Revenue Service rules implemented shortly after Hurricane Katrina struck and recent legislation, hurricane victims may delay all tax payments until February 28, 2006. Taxpayers with records in the disaster areas and relief workers can also qualify. As result, some payments of taxes, including estimated payments of corporate and individual income taxes and withheld income and employment taxes, may be delayed for a number of months. Some of that delay would shift receipts out of fiscal year 2005 and into 2006, but probably not more than a few billion dollars.

In addition, the Katrina Emergency Tax Relief Act of 2005, which was enacted on September 23, provides tax relief in a number of ways to businesses and individuals. The Joint Committee on Taxation estimates that the act will reduce revenues by about $6 billion, almost entirely over fiscal years 2006 and 2007. The provisions with the largest revenue effects allow taxpayers to deduct more personal property losses from taxable income, allow taxpayers more time to replace damaged property without being assessed income taxes on the insurance proceeds, and allow businesses and individuals to deduct more charitable donations from taxable income.

EFFECTS OF THE HURRICANES ON STATE BUDGETS

Data from the state of Louisiana are especially difficult to acquire, but that state is expected to face severe revenue problems. Early information from Mississippi, Alabama, and Texas indicates that state general fund revenues may not suffer significantly as a result of Hurricanes Katrina and Rita. Expenditures are expected to increase significantly in all four of those states, whether from influxes of evacuees in Texas, or from cleanup and rebuilding efforts in Louisiana, Mississippi, and Alabama. Some local governments may confront more difficulties, as they face significant losses in their property tax bases.

Louisiana

State officials are still in the process of gathering information about the effects on the state budget. Unofficial estimates of lost revenues have ranged from $3 billion to $6 billion. The governor's budget recommendation for assumed state tax collections for state fiscal years 2005 and 2006 is about $12 billion, excluding federal funds of over $6 billion. Business closures, disruptions in corporate tax filings, and the mass departure of tens of thousands of residents will disrupt both income and sales tax collections. According to the Government Performance Project, the state relies on sales taxes for almost 60 percent of its total tax collections and income taxes make up about 28 percent. About $2 billion in state sales tax collections comes from the affected areas.

Local governments, particularly New Orleans', are in a more precarious position as they have lost significant parts of their tax bases—notably revenues from property taxes. About two-thirds of the population of Louisiana is from an area officially declared a disaster area. Property taxes in the affected counties total about $1.3 billion, and local sales taxes about $1.8 billion. The affected local governments have about $9 billion in outstanding debt.

Mississippi

In testimony before the state economist's office on September 12, the Deputy State Economist said that the revenue picture would be mixed but that over time, the net effect on the state's general fund would probably end up being negligible. While about two-thirds of the Mississippi population lived in an area officially declared a disaster area, initial tax losses resulting from lost income and wages and some decreased gaming activities will be balanced by increased collections from income taxes as cleanup continues and rebuilding efforts begin.

About one-third of the state's gaming facilities have been affected by Hurricane Katrina, but revenues from gaming support only a small portion of the state's overall general fund—about 5 percent. A little less than half of those revenues comes from the coastal counties most affected by the hurricane.

The state is trying to assess the potential damage to state facilities and any increased demands on state programs. Much of the ultimate effect on the state budget will also depend on how much of the spending is eligible for federal reimbursement.

It is likely that local governments in areas affected by the hurricane were generating between $250 million and $300 million from local sales taxes. With respect to the property tax, the affected areas generate about $2 billion in property taxes, or a little more than two-thirds of the state total, and have about $4 billion in outstanding debt.

Alabama

Although Alabama's gulf coast was hit by Katrina, the state is not anticipating any long-term effect on revenues. The 10 affected counties in Alabama hold about 18 percent of the state population.

The primary effect on state revenues will be any reduction in income taxes as a result of lost wages. At this time, state officials are not anticipating a significant change in sales tax revenues. The state is granting a waiver in the six counties most affected by the hurricane for income tax remittances; individuals and corporations will be able to delay paying income taxes until January 1. Early estimates indicate that the state's cash flow will be sufficiently strong to absorb the effects of that delay.

Local governments in areas affected by the hurricane generate about $243 million in property taxes and $267 million in sales taxes. The affected local governments have about $2 billion in outstanding debt.

Texas

Texas has received a majority of the evacuees from Katrina—more than 140,000 displaced heads of household are registered in Texas with the U.S. government. The state estimates that 373,000 evacuees are temporarily calling Texas home. The state also sustained substantial damage, particularly in rural areas, from Hurricane Rita.

While the governor's office has stated that it is too soon to estimate the total costs of the storms to the state and local agencies, the costs from various agencies are considerable. The Texas Education Commissioner estimated that the state's bill to assist displaced students could be at least $450 million. The city of Austin, which is housing over 4,000 evacuees, estimated costs at $2.7 million in less than two weeks. The city may receive federal reimbursement for some of those costs.

FEMA has said that it will reimburse school districts for transportation costs, portable buildings, school computers, and mental health counselors, but not for other costs related to hiring new teachers and support staff.

EFFECTS ON MUNICIPAL BONDS

Overall Picture

All three major ratings firms have issued warnings on debt issued by cities, special purpose districts, and states along the Gulf Coast, but they had not downgraded any bonds as of September 23. Because of the hurricanes, between $10 billion and $15 billion in outstanding debt in Alabama, Louisiana, and Mississippi is at a greater risk of default. Also, local governments and quasi-governmental entities are facing significant risks of default. In general, most analysts believe that the hurricanes have increased the chances of short-term losses but that long-term stability will return to the market because many of those losses will be covered by insurance and because the federal government will provide significant aid for cleanup costs.

Local and Quasi-Governmental Issues

Both Louisiana and Mississippi have relatively good liquidity and the ability to borrow from various funds in order to meet bond obligations. In contrast, local governments face lost property and sales taxes and will probably not be able to generate other revenue streams to pay off bonds. Although they can probably pay what is owed through the end of the year, local governments will have limited options to raise revenues after that. Some issuers were not able to make September interest payments, and some analysts believe that bankruptcy for New Orleans and other cities is a possibility. MorganStanley suggests that health care facilities and certain structured public finance transactions like those for stadiums and transit projects are at the most risk for default.

Louisiana and Mississippi are working to help local entities pay interest on bonds. In Mississippi, the state has established a $100 million fund from which local governments can borrow and pay off the loans with tax revenues or funds from FEMA whenever they are received.

New Orleans has about $800 million of outstanding bonds, the water and sewer authority another $198 million, the convention center has about $500 million in bonds that are dependent on hotel and restaurant taxes, and the airport has about $140 million in outstanding general airport bonds.

A study by Standard & Poor's suggests that bond insurers with the heaviest exposure in the Gulf Coast region will be able to withstand "worst case" scenarios, remaining liquid even if required make payments on numerous defaults. Moody's is somewhat less optimistic, suggesting that a worst-case scenario would put capital ratios for certain firms at a precarious level, and might generate a downgrade in ratings for bond insurers.

SECTORAL ECONOMIC EFFECTS

Energy Supply and Prices

Supply has been lost in the markets for crude oil, petroleum products (especially gasoline), natural gas, and electricity. Supply that was returning to the market after Hurricane Katrina was further disrupted by Hurricane Rita. The prices of gasoline and natural gas immediately jumped following the disruption of refinery and pipeline activities. Crude oil prices were not greatly affected by the loss of offshore oil production because refinery demand for oil was curtailed and because sales from the U.S. Strategic Petroleum Reserve and by the International Energy Agency had been announced. With offshore production initially returning faster than refineries were restarting, crude oil prices actually fell from their pre-Katrina levels—although renewed increases in crude oil prices are likely if the recovery of offshore production is delayed. Gasoline prices started falling after Katrina, too, with the reopening of major product pipelines and increasing refining activity but remain above pre-Katrina levels. Natural gas prices, however, remain high and are likely to stay that way for some time.

The higher prices attributable to the two storms will linger for some months, until the damages to offshore oil and gas platforms and pipelines and to petroleum refineries, natural gas plants, onshore storage terminals, and transportation infrastructure for oil and gas can be repaired. Upward pressure on electricity rates is likely in the future as utilities try to pass through their increased costs for natural gas and as they press state regulatory agencies to allow them to recover costs for repairing damaged equipment and recover revenues from lost sales.

Crude Oil

Hurricane Katrina resulted in an immediate loss of nearly 1.5 million barrels a day of crude oil from the Gulf (100 percent of Gulf production, and 10 percent of the nation's oil needs), but the loss fell to about 0.9 mmbd before the facility closures in advance of Hurricane Rita halted all production again. About 700 manned platforms for producing oil and gas currently were closed as of September 28.

The full restoration of oil production will depend on the extent of damages to production platforms and undersea pipelines, and the damage assessment is far from complete. (About 0.2 mmbd are still offline because of damage to pipelines that Hurricane Ivan caused a year ago.) After Katrina, the Minerals Management Service reported extensive damage to 20 offshore production facilities and the destruction of 46 low-producing structures. The damaged structures include some of the highly productive deepwater platforms, which may not fully recover for many months. Hurricane Rita may have delayed the work in assessing the damage from Hurricane Katrina as well as inflicted additional damage. The extent of further damage is unknown, but it is likely to be extensive, given the intensity and path of the second storm.

One main factor has so far prevented the production losses from pushing up crude oil prices: refinery shutdowns have reduced the local demand for crude oil. Other potential influences on oil prices that have been reported are the announced sales of strategic oil reserves and the announced increase in OPEC's (the Organization of the Petroleum Exporting Countries') production quotas. Those factors are not likely to have a major impact on the price of crude oil, however.

Gasoline

The loss of gasoline supply was a problem almost exclusively for the Southeast and the East Coast initially as a result of the closure of two major product pipelines, but they soon reopened. Refinery closures attributable to Katrina continue to reduce the nationwide supply of gasoline and other refined products by about 5 percent (about 0.9 million barrels per day). In addition, a significant amount of refinery capacity currently is shut down because of Hurricane Rita. Altogether, 20 percent of capacity is shut down, and it is uncertain how soon those plants will reopen.

In the first week after Katrina, the price of gasoline increased by 60 cents a gallon in the East (or by 25 percent), to nearly $3.20 a gallon (for regular grade). Gasoline prices also increased sharply elsewhere—nationwide, prices averaged about $3.05 a gallon. Outside the East, the price increase largely reflected a fivedollar-a-barrel jump in crude oil prices that occurred in August before the hurricanes and by the normal Labor Day peak in gasoline demand: it does not appear to have been caused by the hurricanes.

The initial loss of gasoline supply to the East was about 30 percent (reflecting a cutoff in the flow of both Texas and Louisiana gasoline) and to the nation, about 10 percent. With the recovery of the product pipelines within a week and with the benefit of a significant inventory drawdown in the East, the total supply to that region after the first week was down by only about 10 percent. Despite the closure of additional refineries with Rita (reflecting 3.5 mmbd in refining capacity), the market has not reacted further. In the week before and the week after Hurricane Rita, nationwide gasoline prices averaged about $2.80 a gallon. However, because capacity will be lower for many months, gasoline prices are likely to be higher during the first half of 2006 than they would have been in the absence of the storms.

Natural Gas

Natural gas supply was affected by the loss of production from the Gulf of Mexico, damage to natural gas plants (which remove certain liquids from the gas so that it can go into pipelines for commercial use), and damage to pipelines. For the most part, the pipeline outages were related to electricity supply problems and to the slowness of the damage assessment work. Damage assessment also depends on the pace of the restoration of natural gas production. (Leaks cannot be found if pipelines are not operating at capacity.) The initial loss of nearly 10 billion cubic feet (bcf) a day from the Gulf, nearly 10 percent of the nation's supply, had been cut to about 3.5 bcf before Rita hit, but the volume that was shut in increased again with that storm. Currently, offshore gas production is about 7.8 bcf a day (or 80 percent of total capacity in the Gulf). Initial indications are that natural gas production will not come back as quickly following Rita as it did following Katrina.

Wholesale prices for natural gas increased from about $9 per thousand cubic feet (mcf) to a high of about $12 per mcf just after Katrina. Unlike gasoline prices, natural gas prices did not come back down as additional supply came online, and with the new loss from Rita, they currently are close to $13 per mcf (having peaked at over $14). Natural gas prices may well stay at those record highs through the winter. If the supply remains seriously disrupted, prices could go up even more. Damage assessment for the offshore regions has barely started.

Sustained high prices are related to the inelasticity of demand for natural gas by electric utilities in the face of the large supply disruption. Reportedly, utilities throughout the South have experienced difficulty getting natural gas. Operators of the Henry Hub (that routes Texas and Louisiana natural gas into the major interstate pipelines) are rationing their supplies. Further, late summer and early fall are normally a time to build inventories for the winter. That buildup now appears to be on schedule, although continued reduced production in the Gulf probably would lower it and contribute to continued high prices through the winter.

Electricity

After Hurricane Katrina, about 2.7 million customers were without electric power, state utilities in the Gulf had restored power to all but about 250,000, but with the further damages from Hurricane Rita, the outages increased again. Three days after Rita, customers without power in Texas, Louisiana, Mississippi, and Arkansas numbered about 1.3 million. Entergy, the main utility in Louisiana and east Texas, reports 279 transmission lines and 284 substations out of service—almost all of that damage is attributable to Rita. (The damage to 181 lines and 263 substations from Katrina had been largely repaired.)

Entergy had reported total losses for damages and lost sales related to Katrina of about $1 billion. After Hurricane Rita, Entergy and the utilities operating in the affected areas of Texas, Louisiana, and Mississippi now report that they will need $2.5 billion in assistance. Those costs have to be recouped through some combination of insurance, diminished returns to investors, and increased rates for local electricity consumers—subject to the approval of state public utility commissions. More broadly, rates for electricity in the coming months also will reflect a pass-through of increased costs for natural gas—a consequence of damage to offshore production and to pipeline infrastructure.

Transportation Sector

The hurricanes do not appear to have caused major disruptions to the transportation system of the United States—the only damage that appears likely to cause significant problems is that to the Port of New Orleans. Although some highways, rail lines, ports, and channels sustained severe damage and will operate at levels substantially below their prehurricane capacity for weeks or months, redundancies in the transportation system will enable most shipments to reach their destinations, albeit at somewhat higher costs and times in transit.

Roads and Bridges

In the immediate aftermath of Katrina, roads throughout southern Mississippi and the New Orleans area were impassable because of debris and downed power lines. Within a few days, however, most roads had been cleared enough to permit travel, although in some places, only two lanes of four-lane highways were available. Hurricane Rita flooded some roadways, but major arteries were not affected.

At this point, the national effects of road damage appear minimal, because most major roads reopened quickly and because alternative routes beyond the hurricane zone could accommodate rerouted traffic. Highways into and out of the ravaged areas of southern Mississippi and Louisiana have not been put to the test, however, because diminished economic activity in those areas has meant less demand for truck transportation (the major exception being the influx of emergency supplies).

Local and regional effects are greater. The I-10 Twin Spans Bridge across Lake Pontchartrain between New Orleans and Slidell needs its deck replaced. Louisiana highway officials expect to restore one lane in each direction by October 26, 2005, and two lanes in each direction by January 6, 2006. Because I-10 is the most direct route from the Port of New Orleans to points in the East and Northeast, the loss of it will increase trucking time and costs. But a work-around to the west and north is available. Once the port returns to normal, the cost and increased time associated with the diminished capacity of the I-10 bridge will probably be more pronounced. Numerous bridges on U.S. 90 and on state and local roads in southern Louisiana sustained serious damage. Many of those bridges provide access to port terminals and other commercial facilities as well as serve local traffic.

In Mississippi, two bridges on U.S. Route 90 (which parallels the east-west I-10) were destroyed: the Bay St. Louis Bridge and the bridge between Biloxi and Ocean Springs. In addition, the road was badly damaged in many places. The Mississippi Department of Transportation expects to have two lanes of traffic on U.S. 90 in Harrison County open for civilian traffic by mid-November.

Freight Railroads

Railroad lines in the New Orleans area and to the northeast suffered extensive damage, and a substantial amount of traffic that ordinarily would be interchanged between eastern and western railroads in New Orleans continues to be rerouted by way of Memphis or St. Louis. That adds as much as 48 hours to transit time. The most badly damaged track appears to be CSX's 100- mile line between New Orleans and Pascagoula, Mississippi. When it will be back in full service is not known. The consequences of that damage will be more apparent once port operations return to normal levels.

Ports

Several major ports along the lower Mississippi River in Louisiana and on the Gulf Coast sustained damage.

Of the Mississippi River ports—the Port of New Orleans, the Port of South Louisiana, and the Port of Greater Baton Rouge—New Orleans sustained the most serious damage but is slowly resuming business. It expects to be operating at about 30 percent to 40 percent of capacity within a month, at 80 percent of capacity within three months, and at 100 percent in four to five months. The first cargo since Katrina hit was unloaded on September 13. As of that date, trucks were able to get in and out of the Port of New Orleans.

Those ports handle large amounts of bulk cargoes, such as crude and refined petroleum products, steel, aluminum ores, chemicals, rubber, machinery, plywood, coffee, cotton, foodstuffs, forest products, corn, oil seeds, and wheat. Ranked by tonnage, the Port of South Louisiana was first in 2003, New Orleans was fifth, and Baton Rouge was 10th. The Port of South Louisiana handles half of all U.S. exported grain and grain by-products, and its major test will come over the next few weeks as grain is harvested in the upper Midwest.

Each port has numerous terminals and extends many miles along the river. Many of the terminals concentrate on particular types of commodities for which they have specialized equipment. Some appear to have suffered greater damage than others, making the overall effects on commerce unclear. An underlying problem common to the Port of New Orleans and the Port of South Louisiana has been the lack of electricity and a lack of workers. Efforts are under way to restore power, and the Maritime Administration is working on getting ships in place that could house workers temporarily.

The Louisiana ports handle a smaller percentage of the nation's containerized cargo. Because the value of such cargo, which includes items like electronics, toys, clothes, and other household items, is generally much higher per ton than that of bulk commodities, the value of traffic handled at the Louisiana ports lags far behind that at the giant ports of Los Angeles, New York, and Long Beach.

Along the Gulf Coast in the state of Mississippi, the ports of Gulfport and Pascagoula sustained extensive damage. It is unknown when they will be able to resume normal operations. In the meantime, ships are being diverted to other ports. In 2004, Pascagoula ranked 35 and Gulfport ranked 37 in terms of the dollar value of foreign trade at U.S. ports, with each handling about $4 billion of cargo.

In Louisiana, Port Fourchon, a key facility for crude oil transportation, has suffered major damage.

Waterway Navigation

The Mississippi River is now open to deep-draft navigation along its entire length, although a part of the southern end—roughly the portion encompassing the Port of New Orleans—is limited to daylight operation because some aids to navigation have not yet been replaced. The Coast Guard is advising vessels to operate at their minimum safe speed. Should the combination of slow orders and delays at terminals persist, barges will be unable to cycle up and down the Mississippi as quickly as usual, and shippers' costs could increase. Along various canals, sunken or damaged barges that might pose hazards to navigation are being identified and removed.

Drinking Water and Sewage Treatment

Hurricane Katrina caused significant disruption to drinking water and sewage treatment operations, and Hurricane Rita caused additional damage. Full restoration of drinking water systems and wastewater treatment systems will be delayed by the many breaks in the distribution and collection systems and by the need for upgrades and repairs in older systems. Uncertainties arising from the need to drain the floodwaters and from the scope and complexity of the interdependent systems make it difficult to estimate recovery times.

As of September 25, with the impacts of Hurricane Rita awaiting assessment, the Environmental Protection Agency (EPA) identified 158 drinking water systems and 40 publicly owned wastewater systems in Louisiana, Mississippi, and Alabama that were not operating or whose status was unknown. Those figures represent 7 percent of the drinking water systems and 16 percent of the publicly owned wastewater systems in the affected areas of those states. Systems considered operational may also need significant repair or restoration.

Estimates of the ultimate cost to restore drinking water and wastewater services are preliminary and vary widely. EPA does not yet have access to all of the affected systems and does not expect to have cost estimates for weeks.

Although water systems will continue to rely in part on the EPA-administered state revolving funds (SRFs) for clean water and for safe drinking water, FEMA grants will be the most immediate source of funds used to rebuild systems. The amounts currently available through the SRFs are small relative to the projected need. For example, allocations to the clean water SRFs of the three Gulf Coast states affected by Katrina totaled $41.3 million in fiscal year 2005 and are tentatively set at $28.6 million for fiscal year 2006. Those states' safe drinking water SRFs received a total of $24.9 million in fiscal year 2005 and are tentatively allocated $28.6 million for fiscal year 2006. FEMA's public assistance program grants typically constitute two-thirds of federal moneys flowing to disaster areas.[3] The grants (authorized under the Stafford Act) are classified as for either emergency work (that which must be done immediately to save lives, protect property, public health and safety, or to lessen the threat of a major disaster) or permanent work (intended to restore damaged facilities to their predisaster design). Categories of permanent work, which may extend for 48 months, include water treatment and delivery systems and sewage collection and treatment facilities.

Other federal programs can also provide postdisaster assistance to water systems. The Department of Agriculture's Rural Utilities Service offers emergency and imminent community water assistance grants to respond to significant actual or imminent declines in the quality or quantity of drinking water where funds are needed to meet health or sanitary standards such as Safe Drinking Water Act and Clean Water Act requirements. EPA's Drinking Water Emergency Assistance program (authorized under the Safe Drinking Water Act) provides technical assistance and grants to respond to emergency situations in which public water systems, including their water sources, present a substantial danger to the public health as determined by the Administrator. Finally, under the National Response Plan administered by the Department of Homeland Security, the Army Corps of Engineers is responsible for FEMA's emergency support for public works and engineering. Such support includes emergency repair of wastewater facilities and temporary restoration of water supply systems. Funding is provided through mission assignments made by FEMA and paid for from the Disaster Relief Fund.

Bank and Other Loan Repayments

Default rates on business loans, as well as residential and commercial mortgages, are likely to increase, but insurance payments may keep the default rates low. In many cases, the value of the collateral backing the loans has been greatly diminished. However, the prospects for banks in the area might actually be improved by the increase in lending that will accompany rebuilding. Banks in Florida did well after the four hurricanes that struck last year. What is uncertain is whether that experience is a reliable guide because the nature of the losses is different.

Standard & Poor's expects the effect on residential mortgage-backed securities (RMBSs) to be "relatively limited." (The RMBS transactions do not include agency mortgage-backed securities; those transactions would largely be the "private-label" market.) The exposed pools have a limited concentration of mortgages in the affected areas. Fitch Ratings reached a similar conclusion after a review of two-thirds of its rated RMBS portfolio. Of the $475 billion reviewed by Fitch, just $3 billion (0.65 percent) are secured by properties in FEMA-designated disaster areas. Fitch believes that uncompensated losses on that exposure will not be substantial.

Most losses will probably be associated with the lack of flood insurance or insufficient flood insurance. Fitch's review found that the servicers of the mortgages were required to maintain flood insurance in special flood hazard areas, and to attest to that coverage in the servicer representations and warranties. However, some parts of New Orleans were not designated as special flood hazard areas. Initial indications are that between 25 percent and 40 percent of homeowners in New Orleans have flood insurance. Those without insurance will probably receive FEMA or SBA assistance in the form of cash grants or low-interest loans. In addition, some borrowers may face extended unemployment, and others may choose not to return. Standard & Poor's expects that there will be an increase in delinquencies, given the damage to homes and the loss of employment.

In the past, hurricanes and flooding have not materially affected the trend of Freddie Mac's credit losses. Freddie Mac's exposure in the FEMA-designated disaster areas represents less than 1 percent of its total $1.2 trillion credit guarantee portfolio, with just a quarter of that exposure in the three hardest hit parishes in the New Orleans area.

Agriculture Sector

Winds during Hurricane Katrina affected crops in portions of Alabama, Florida, Louisiana, Mississippi, and Tennessee. The affected counties account for 1 percent to 3 percent of total U.S. production of hogs, milk cows, cattle, corn, and soybeans. Those counties also account for around 15 percent of broilers, upland cotton, and rice and 85 percent of U.S. sugarcane.

Estimates of agricultural losses resulting from Katrina are still in flux, and estimates for Rita are not yet available. On September 20, 2005, the Agriculture Department estimated that losses due to Katrina amounted to $700 million for crops and $170 million for livestock. In addition, about 4.2 billion cubic feet of timber was damaged, and net losses will depend on the success of salvage operations. Katrina also caused significant damage to livestock production facilities. An estimated 6.2 million chickens died in the storm, and 2,400 poultry barns in the region were either lost or damaged. Loss of power to dairy operations and impassable roads have made it difficult to process and transport milk to market. Dairy losses are around $3 million per week as long as power shortages continue. Those estimates could change as formal damage assessments are completed.

The agriculture sector will also be affected by higher prices for gasoline, diesel, and fertilizer. For each month that the cost of fuel remains 10 percent higher than it was before the storm, farmers incur additional production expenses of about $85 million. The domestic nitrogen fertilizer cost index is currently 13 percent above what it was in August of last year.

Depending on the severity of the losses and the level of insurance coverage purchased, producers incurring crop losses from Katrina may be partially reimbursed through the federal crop insurance program. No estimates of insurance payments resulting from the storm are available at this time.

Although Katrina damaged port facilities, and a few facilities may takes months to become fully operational, grain shipping is not expected to be a major problem. A current shortage of barges for Gulf transport and higher energy costs have increased the cost of moving grain, resulting in somewhat lower crop prices upriver in the major grain belt areas. But port facilities are expected to be able to handle this year's grain harvest, which will be in full swing in a matter of weeks.

On September 21, the Agriculture Department announced administrative measures to alleviate stress on the grain economy caused by Katrina's damage to the Gulf transportation system. Those actions include providing financial incentives to move barges of damaged corn from New Orleans; arrange for alternative storage for grain; find alternative shipping routes to relieve pressure on the Gulf system; and allow producers to store government-owned corn on the farm with the option to purchase it.

End Notes

[1] See www.rms.com.

[2] Those data are derived from reports submitted by all employers who are subject to unemployment insurance, accounting for more than 98 percent of total wage and salary employment. The Quarterly Census of Employment and Wages is sometimes referred to as the ES-202 program.

[3] The Congress appropriates funds for the program annually to the Disaster Relief Fund. Supplemental appropriations have also been made in response to specific disasters.

INDEX

C

D

E

F

G

H

I

L

S

T

U

V

W